Technology-Based
LEARNING

Maximizing Human Performance and Corporate Success

Michael J. Marquardt
Global Learning Associates
Reston, Virginia

Greg Kearsley
Southeastern Nova University
North Miami Beach, Florida

ASTD American Society for
Training & Development

CRC Press
Taylor & Francis Group
Boca Raton London New York

CRC Press is an imprint of the
Taylor & Francis Group, an **informa** business

CRC Press
Taylor & Francis Group
6000 Broken Sound Parkway NW, Suite 300
Boca Raton, FL 33487-2742

ISBN-13: 978-1-57444-214-4 (hbk)
Library of Congress Card Number 98-23195

Library of Congress Cataloging-in-Publication Data

Marquardt, Michael J.
 Technology-based learning : maximizing human performance and corporate success / Michael J. Marquardt, Greg Kearsley.
  p. cm.
  Includes bibliographical references and index.
  ISBN 1-57444-214-7 (alk. paper)
  1. Employees—Training of. I. Kearsley, Greg, 1951- . II.Title.
HF5549.5.T7M286 1998
658.3'124—dc21   98-23195
                                                                  CIP

Visit the Taylor & Francis Web site at
http://www.taylorandfrancis.com

and the CRC Press Web site at
http://www.crcpress.com

About the Authors

Dr. Michael J. Marquardt is a professor of Global Human Resource Development at George Washington University and President of Global Learning Associates. He has held a number of senior management and training positions with organizations such as Grolier, TradeTech, Association Management, and the World Center for Training and Development. Mike is the author of more than 60 books and articles in the areas of learning, technology, and globalization. He serves as a senior advisor to a number of Fortune 500 firms as well as to government agencies in the U.S. and overseas.

Dr. Greg Kearsley is one of the pioneers in learning technologies. He has served as Chief Executive Officer of Park Row, Inc., a software publishing company, as a chief scientist at Courseware Incorporated, and as a senior engineer at HumRRO. An author of numerous books on technology and education, Dr. Kearsley has taught at Nova Southeastern University, the University of Wisconsin, the University of San Francisco, and San Diego University.

Contents

v

Preface

Technology has dramatically changed the way we learn, the way we work, the way we live, even the way we think. It has likewise fundamentally altered the way organizations are structured, the purpose of management, and the manner in which companies relate to their customers. Technology has totally transformed the system in how employees are educated to carry out their work, prompting radical new questions such as — Should there be training at all, or do we now just provide people with the tools and resources to access knowledge?

This book explores these critical questions as well as the interplay and growing interdependence among the forces of *technology, learning,* and the *workplace.* Each of these forces is undergoing incredible and astonishing transformations as we enter the new millennium.

Technology now integrates television, telecommunications, and computers through digitization and compression techniques. Neural networks can create associative reasoning and can process complex questions. Superconducting transmission lines can carry one trillion bits of information a second, enough to send the complete contents of the Library of Congress in two minutes! Internet usage has gone from fewer than 1000 users 15 years ago to over 100 million today. Biotechnology will soon enable people to walk around with microchips in their bodies that will monitor and assist their day-to-day activities.

Technology also accelerates global competition and worldwide communications. It does much of the work we used to do; i.e., manufacturing (using our hands) so that we can do the more needed work of mentofacturing (using our minds). It has enhanced workers' power to learn. The increased and intelligent use of technology for education and training has become fundamental for the continuing economic development of organizations and, indeed, of nations.

Learning has become a lifelong process. For thousands of years the traditional way of education and learning was to bring students together in a single location (usually academic) for lectures that occurred in one's early years. Now technology and competitive pressures have made learning all-the-time, everywhere, yet just-in-time and customized just-for you. We must now train ourselves (i.e., take the responsibility and develop the competencies of self-directed learning).

The location of learning is shifting from schools to the workplace. McDonald's trains more adults a year (700,000) than the California university system. Employee education is growing 100 times faster than schools. Shoshanna Zuboff, in her classic *In the Age of the Smart Machine*, foresaw the impact of technology in the workplace and made the paradoxical statement that "learning is the new form of work."

Learning has become the most important part of everyone's job. The continuously changing knowledge requirements of workers create a continuous need for learning, learning so voluminous that it can only be done with the wise use of technology. Although technology has the power to speed up our learning and make knowledge more accessible, it is still people who have to have the desire and skills to access this knowledge. And the ability to access "data" and convert it into "knowledge" has become a prime requirement and skill of all workers.

Workplaces too are rapidly changing. For a growing number of us, the workplace is our home or car. Many of us work for many organizations. We are closer to our customers than to the people across the hall. Managers manage knowledge more than people. And more and more of our work involves knowledge and technology.

Organizations have gone from the industrial era to the service era to the knowledge era within the past 50 years. Now companies are moving from the quality efforts of the 1980s through the reengineering processes of the 1990s to the radical transformation of the workplace itself as we enter the 21st century. Workplaces have moved from focusing on the reduction of defects and the streamlining of business processes to totally new forms that enable organizations to manage continuous, white-water change and urgently strive to become learning organizations. Technology allows for the creation of high-performance workplaces in which work is being reorganized, redesigned, and reengineered to improve performance.

Purpose of This Book

The old, traditional ways of learning and working in the workplace have become irrelevant, both in their methods and in their very purposes, because of the power of technology. This book explores the interplay and impact of technology on how we work and learn, how we manage ourselves in the workplace, and how we manage knowledge in our lives. It seeks to show how technology can help you and your organizations achieve the level of competence necessary to survive and succeed in the global marketplace of the 21st century.

Writing This Book

Technology also changes the way books can be written. The two co-authors live in different parts of the U.S. and frequently travel to Asia, Europe, and Africa. Writing this book was accomplished primarily via the Internet. Correspondence was by e-mail. We even created a web page for the book (http://www.gwu.edu/~lto) to invite comments and ideas on our topic from the general public. (An individual approached one of the authors at a global conference in Kuala Lumpur to indicate that he recognized his name and activities from visiting this web page.) Forums to discuss various chapters of the book are available through the web page. Much of the research was done searching the Internet and using the reference services found on the world wide web. We welcome readers' comments on this book sent to our e-mail addresses below to guide us in future editions of this book.

Overview of Book

Part I describes the rapid changes in technology and the emerging impact of that technology on the world of work (Chapter 1), on the workplace (Chapter 2), on the organization, especially in building a learning organization (Chapter 3), and on the worker (Chapter 4). We will see how all of these entities have been dramatically altered because of technology, such that learning and work have become synonymous, that the workplace can be anywhere, and that organizations cannot survive without continuous learning and ubiquitous technology.

In Part II, we begin by providing an overview of electronic and computer-based technologies that turbocharge the speed of learning, allow training to be anywhere, anytime and for anyone, and dramatically change the role of trainers and managers (Chapter 5) We then present six key learning technologies beginning with electronic publishing (Chapter 6). We move to exploring television and video (Chapter 7), teleconferencing by audio, video, and or desktop (Chapter 8), interactive multimedia (Chapter 9), simulation and virtual reality (Chapter 10), and conclude with authoring (Chapter 11).

In Part III, we explore how technology is being used to manage knowledge and provide just-in-time information and skills to the worker. Chapter 12 describes how technology can be used to acquire, store, analyze, transfer, and apply knowledge. Following a discussion of electronic performance support systems in Chapter 13 and the networks (Internet, intranet, world wide web, WANs, and LANs) in Chapter 14, we conclude this section with a description of knowledge engineering by noted guest author, Doug Weidner (Chapter 15).

The final section of the book includes principles and practices necessary to successfully utilize technology in the global workplace. Chapter 16 guides the reader in the process of selecting and evaluating various technologies for the workplace. In Chapter 17 we look at how culture and globalization affects the use of technology. Chapters 18 and 19 provide seven case studies of best practices in learning technologies and knowledge-management technologies. Chapter 20 looks at future prospects for technology in the world of work.

Welcome to the Technological Workplace and Learnplace

Much of today's working and learning can only be accomplished with the support and guidance of technology. Technology does not replace workers, but dramatically enhances their opportunities and powers to serve customers, fellow workers and themselves. These are truly exciting times for being in a workplace, which is increasingly becoming a place of learning. We should all seek to become more competent and comfortable in this new world which merges technology, learning, and work. We hope this book will help you appreciate and enjoy this adventure!

Michael Marquardt
mjmq@aol.com
Reston, Virginia

Greg Kearsley
kearsley@fcae.acast.nova.edu
North Miami Beach, Florida

May 1998

Acknowledgments

Many people have been instrumental in helping us write this book. First of all, Drew Gierman at St. Lucie Press and Nancy Olson of the American Society for Training and Development who recognized the importance of a book that combined the elements of technology, learning, knowledge management, and the workplace.

We are grateful for the many people who contributed to the book by actually writing sections, offering advice, providing ideas, and allowing us to research their organizations: namely, Charles Tweedly of Otis Elevators, Tom Solomon of Ernst & Young, Larry Conley of Ford Motor Company, Gloria Gery of Gery Associates, Lew Parks of AMS, Adele Ewing of Government Accounting Office, Andrew Gibbons of Utah State University, Debbi Steinbacher of Price Waterhouse, Angus Reynolds of Southern Illinois University, Deborah Masten of JCPenney, John Didier and Edith Pena of the World Bank, and Randy Maxwell of Nortel.

Finally, we want to dedicate this book to our families for their continuous encouragement and support for our writing efforts. Thanks for the "high-love" to complement the "high tech."

1 Explosion of Technology in the World of Work

Welcome to the new technological workplace with teletraining, info-structures, and ubiquitous computers! Alvin Toffler writes how the advanced global economy and workplace cannot run for thirty seconds without the technology of computers. Yet, today's best computers and CAD/CAM systems will be "stone-age" primitive within a few years. The workplace will demand and require ever more technological advancements and innovations.

Already we have technologies such as optoelectronics, DVDs (digital video discs), information highways, LANs (local area networks), WANs (wide area networks), groupware, virtual reality, and electronic classrooms. The power of workplace computer technology has progressed from mainframe to desktop to briefcase portable to the user's hand. More and more of a company's operations require computer-generated automation and customization.

These technologies have become necessary to manage the "data deluge" so that we can learn faster in fast-changing, turbocharged organizations. Working in a global economy in which "being informed, being in touch, and being there first" can make all the difference between success and second-best, technology provides a big advantage indeed!

Technology Transforms the Workplace and Learning

The impact of technology on organizations, on management, and on learning is mindboggling. And it has only begun. The emerging power and applicability of technology will turn the world of work on its head. Organizations will become more virtual rather than physical because of technology. People will be more closely linked to customers in Kuala Lumpur than to co-workers across the hall because of technology. Technology will cause learning to become the prime purpose of business rather than work itself, and learning, as Zuboff proclaims, will become "the new form of labor."

Technology will increasingly require that managers manage knowledge rather than people. Technology will alter how and why workers learn. Employees will need to train themselves (i.e., learn), and workplace learning will no longer be in a fixed time and location with a group of people for just-in-case purposes; instead, it will be implemented on a just-what's-needed, just-in-time, and just-where-it's-needed basis. The technological forces that have already restructured work will force those responsible for employee development to "create ever more flexible and responsive learning and performance solutions" (Bassi et al.).

To better prepare us for how technology will transform work, workers, and the workplace, it is important to grasp some of the emerging ideas and applications of technologies. This understanding can help us redirect that technology and thereby increase the speed of learning and the management of knowledge in the workplace.

With Technology, the Future Is Already Here!

Already, we are living in a world where virtual reality and interactive multimedia technologies will be commonplace. Personalized intelligent agents will soon be available as built-in, on-line experts looking over one's shoulder. Artificial intelligence technologies (expert/knowledge-based systems, speech and natural language understanding user interfaces, sensory perception, and knowledge-based simulation) will be commonly available. Intelligent tutoring systems will be used to allow learner-based, self-paced instruction. Personalized digitized assistants, telecommunications and network advances, groupware, desktop videoconferencing and collaborative software/group systems technology will be prevalent in the next 5 years.

And the speed and impact of technology continues to accelerate! Trying to figure out the capabilities and future directions of this rapidly changing technology is impossible. Let's look at just a few of the already existing powers of technology:

■ Superconducting transmission lines can transmit data up to 100 times faster than today's fiber optical networks. One line can carry one trillion bits of information a second, enough to send the complete contents to the Library of Congress in two minutes.

■ "Neural networks" bring advances in computer intelligence that makes process commands sequentially; this neural network uses associative "reasoning" to store information in patterned connections, allowing it to process complex questions through its own logic.

■ Expert systems, a subset of artificial intelligence, are beginning to solve problems in much the same way as human experts.

■ Telephones are being manufactured that are small enough to wear as earrings.

■ Highly reliable connectivity works regardless of time or place and is easy to use and affordable around the world.

■ Cellular phones can now respond to e-mail.

One of the most amazing and transforming technological additions to our lives is the Internet. The use of the Internet is one of the fastest-growing phenomena the business world has ever seen, building from a base of fewer than a 1000 connected computers in the early 1980s, to over 10 million host computers today.

Intranets (in-company Internets) are rapidly catching up. The implementation of intranets is growing three times faster than that of electronic commercial applications, with over 70% of major corporations currently having or planning intranet application. As the evolution of intranet sites continues, more and more features will emerge. For example, real-time training that combines a live mediator, on-line information, and several remote attendees is already possible. By the end of 1997, nearly 90% of all business had recognized the criticalness of developing comprehensive strategies for using both intranets and the Internet.

Some of the new high-tech learning machines have been called "the most powerful learning tool since the invention of the book." With virtual reality, the mind is cut off from outside distractions, and one's attention becomes focused on the powerful sensory stimulation (light-sound matrix) that bombards the imagination. It becomes possible for ideas and mental images to float in and out of a person's consciousness.

Technology is becoming more and more a part of all products and the total GNP, including aerospace, advanced industrial systems, and automotive.

Already, nearly 20% of an automobile's value is the electronics within it. The computer service and computer software market has grown to over $420 billion, an increase of 50% in the last 4 years! Information technology is expected to form the basis of many of the most important products, services, and processes of the future.

In addition, an array of technological developments has recently emerged for use in the home as well as the office, including:

- integration of television, telecommunications, and computers through digitization and compression techniques
- reduced costs and more flexible use and application of telecommunications through developments such as ISDN, fiber optics, and cellular radio
- miniaturization (tiny cameras, microphones, small, high-resolution display screens)
- increased portability through the use of radio communications and miniaturization
- expanded processing power, through new micro-chip development and advanced software
- more powerful and user-friendly command and software tools, making it much easier for users to create and communicate their own materials (Bates)

The commoditization of ultra-high technology opens spellbinding opportunities for new knowledge-exchange products. British Telecom thinks future generations of portable phones could be installed right in your ear. While talking, the user could also glimpse images or data that are pulled invisibly off the Internet and projected onto a magnifying mirror positioned beside one eye.

The technology of the future will respond to our voices and extend our senses. It will stimulate complex phenomena — weather patterns, stock market crashes, environmental hazards — solving problems and predicting outcomes at a price anyone can afford. Computers — or networks of them — will become ubiquitous as they are invisibly imbedded in other things. These machines will reconfigure themselves when new applications are required. A whole new metaphor for computing is taking shape, patterned on the natural resilience and elegance of biological organisms. They will learn to diagnose, repair, and even replicate themselves (Gross).

"The pace of change is actually accelerating now," says Richard Howard, director of wireless research at Lucent Technologies. In the next two decades we'll see explosive growth of communications, computing, memory, wireless, and broadband technology.

The technology is moving beyond information technology to the "new" area of biotechnology. Biotechnology is widely forecast as being the dominant force of the 21st century. Bioengineered products and biomanufacturing will fuel the new bioeconomy. Major advances in the genetics of DNA will lead to much longer, more productive lives. People will soon be walking around with chips in their bodies that will monitor and assist activities.

Some of the new possibilities on the horizon include a so-called retinal display that "paints" pictures directly on the eye by modulating a stream of photons from light-emitting diodes and scanning them across the retina. The mind perceives these scans as vibrant color pictures. British Telecom's "homo cyberneticus, "for example, shows off its artificial retina and a pacemaker that sends warning signals to the doctor, as well as a vest that turns body heat into electricity.

From Tyrannosaurus *Rex* to "*Rex*" the Computer

Organizations that have not adapted to a world of technology and rapid learning are seen as dinosaurs who will quickly become extinct because they cannot adapt to the new global environment. Starfish Software has recently introduced "Rex," the smallest, coolest computer around, according to the *Wall Street Journal's* Walter Mossberg. The size of a credit card and just 1.4 ounces in weight, Rex can easily fit in a shirt pocket and can store up to 2,500 items — such as names and addresses, appointments, memos — and display them on a sharp, high-contrast little screen. The Rex can connect to a PC, synchronize with programs such as Lotus Organizer or Microsoft Outlook, and suck in all your calendar and contacts data so you can carry the information with you. Push a button and Rex will show you your future (calendar, that is).

Impact of Technology on the New Economy and the New Organization

The global economy, the reengineered organization, and the new-knowledge worker impact technology and are, in turn, profoundly impacted by technology. They are all inextricably linked, enabling one another and yet being driven by each other. Tapscott, in his classic, *The Digital Economy*, has identified 12 different arenas in which technology has impacted the economy and organizational life:

Table 1.1   12 Arenas Impacted by Technology

1. Knowledge economy and knowledge organizations
2. Digitization of communications and services in economy and organizations
3. Virtualization
4. Mass to molecular forms
5. Integration and internetworking
6. Disintermediation
7. Convergence of computing, communications, and content
8. Innovation-based economy and workplace
9. Prosumption
10. Immediacy
11. Globalization
12. Discordance

Source: Tapscott, D. (1995). *The Digital Economy.* New York: McGraw-Hill.

1. Knowledge Economy and Knowledge Organization

The economy has shifted from "brawn to brain." Knowledge is rapidly becoming more and more of every nation's GNP. The knowledge content of products has resulted in smart clothes, smart cards, smart houses, smart roads, and smart cars. In organizations, knowledge work and knowledge workers have become the basis of value, revenue, and profit. Knowledge-management technologies have enabled organizations to learn and the economy to prosper.

2. Digitization of Communications and Services in Economy and Organizations

Communications, services, and business transactions have shifted from analog (physical memos, reports, photocopiers, tape recorders, etc.) to digital (e-mail, multimedia CD-ROM, DVD, etc.) which allows for vast amounts of information to be compressed and transmitted at the speed of light and with excellent quality. Many different forms of information can be combined and thereby create multimedia documents. Information can be stored and retrieved instantly from around the world from a device that fits in your pocket.

3. Virtualization of Economy and Organizations

Physical things can become virtual (organizations, office, village, job, store, even sex) via interactive media or Internet synchronously (same time) or asynchronously (at different times) by means of technology's ability to visualize data, produce real-time animation, and create virtual reality.

4. Mass to Molecular Forms

The industrial economy and structures are being replaced by a molecular economy and structure. We are moving from mass forms of production, media, and government to empowered smaller units of individuals and teams who can access and interact through millions of channels when they choose to do so — thanks to object-oriented systems, technologies, and software that can create products or services for the individual.

5. Integration and Internetworking

The new internetworked economy has deep, yet interchangeable, interconnections within and between organizations and institutions. Networked computing enables the integration of modular, independent organizational components. We are able to link people and information with immense speed and ease.

6. Disintermediation

Elimination of intermediaries that stand between the producer and the consumers (e.g., agents, wholesalers, retailers, middle managers, brokers) is possible because of the shift from multilevel, hierarchical computing architectures to peer-to-peer network computing. More and more of our purchasing will be via interactive television and the Internet, including the purchase of big-ticket items such as automobiles.

7. Convergence of Computing, Communications, and Content

In the old economy, the automotive industry was dominant. The fastest growing sector of the new economy is the new media, which are products of the convergence of computing, communications, and content industries. These new media — such as computer software and telecommunications — are transforming the way we live, work, learn, and even think.

8. Innovation-Based Economy and Workplace

Creativity and human imagination are key drivers of the economy, social life, and the workplace. Companies like Rubbermaid, Motorola, and 3-M recognize that their survival depends on constant innovation. Multimedia technology is a product as well as a driver of innovation. Technology has provided tools to rekindle and stimulate the creativity inherent in all of us.

9. Prosumption

Tapscott notes that the gap between *producers* and con*sumption* has become ever more blurred. Consumers become involved in product design and marketing. Consumers of information and technology become producers; users become designers as they create (author) new software applications themselves. Technology enables users to create systems and databases, to create voice entry and response to multimedia.

10. Immediacy

The new economy is *real-time* and the new organization is *real-time*, both of which are continuously and immediately adjusting to changing business conditions and using technology to communicate at the speed of light, thereby allowing citizens and customers to interact in *real-time*. Just-in-time inventories are the norm, as is just-in-time learning.

11. Globalization

We are part of a global economy. Jack Welch, CEO of General Electric, warned several years ago that "we either globalize, or we die." Global telecommunications enables companies to operate across geographic and time boundaries, to both centralize and decentralize decision-making, processes, and functions.

Technology allows marketing and production to have a global reach but a local touch.

12. Discordance

Just as there is a great divergence between wealthy and poor countries, as well as between rich and poor people, so there is a wide divergence in the availability and capability of using technology among organizations and workers within a locality and around the world. Technology has the power to bring everyone into the global economy and modern workplace, but those without access to it or capability in it will face a widening economic and social gap.

Importance of Technology for Corporate Success

The increased and intelligent use of technology has become critical for the continuing economic development of organizations. Stewart notes that information and multimedia technology, from CD-ROM to CD-I to networked programs and computer-based applications, has become in today's world what the railroad was to 19th century America. Technology utilization is truly the power tool for building the future for today's organizations.

Despite the tremendous importance of technology and the many opportunities for information technology workers, a tremendous shortage of IT workers exists (even with very high average wages of up to $100/hr). The U.S. Department of Commerce predicts that a million new computer scientists and engineers, systems analysts, and computer programmers will be needed in the U.S. in the next 8 to 10 years, nearly doubling the number of workers in these professions in 1998.

A survey conducted by Deloitte & Touche of 1500 chief information officers in 21 countries confirms that the U.S. is not alone in facing these shortages. The global picture is one in which information technology managers around the world are experiencing the difficult combination of "unprecedented demand for IT workers and high turnover rates." Companies are finding it especially difficult to retain employees in four key areas — client/server architecture, data modeling, distributed databases, and packaged software applications. The November 30, 1997, *Washington Post* postulates that the Washington, D.C. metropolitan area has a shortage of over

25,000 technology workers that is costing the local economy over $1 billion a year (Chandrasekaran).

The U.S. Office of Technology Policy declares that, "information technologies are the *most important enabling technologies in the economy today.* They affect every sector and industry in the U.S. ... Severe shortages of workers who can apply and use information technologies could undermine U.S. innovation, productivity, and competitiveness in world markers." Over 50% of company executives cite lack of skilled and trained information technology workers as a *significant* barrier to their companies' growth.

An increasing number of organizations across the world are thus beginning to recognize that they must not only integrate technology into work, but also integrate technology into the growing need for learning.

Increased Demand for Learning in the Workplace

The growing need for workplace learning has occurred because of the changes caused by technology and the tremendous increase in global competition. Providing workers with the knowledge and skills necessary to understand and to compete has become staggering.

The more sophisticated machinery and work processes created by technology require more sophisticated workers. More than half of the new jobs created between 1984 and 2005 will require some education beyond high school. Over 50% of workers now use computers in their jobs. According to a 1996 ASTD survey, 73% of all employees said that computer skills were essential for employment.

Thus, many of the new jobs will require a much higher level of skill than the jobs they are replacing, especially in manufacturing and resource-based industries. People will retain existing jobs only if they are retrained to higher standards.

Learning has been steadily shifting from formal education in the classroom to the factory floors and corporate training centers of the workplace. Employee education is growing at phenomenal rates, such that in recent years, corporate training has increased at 30 times the rate of college education. Academic institutions alone cannot possibly provide for the continuously burgeoning learning requirements of workers.

As a result, corporate universities are springing up throughout the world, attracting world-class researchers as well as gifted instructors. McDonald's Hamburger University alone trains 700,000 people per year.

Knowledge Economy

Technology and globalization has led to a global economy based on knowledge. Knowledge workers now outnumber industrial workers by 3 to 1. The work force has moved from manufacturing (working with the hand) to mentofacturing (working with the mind). Continuous learning and knowledge provide the key raw materials for wealth creation and have become the fountain of organizational and personal power.

The wealth of nations will depend increasingly on knowledge-based, high-tech industries in areas such as biotechnology, health, environmental products and services, tourism and hospitality, telecommunications, computer software and software applications, financial services, and entertainment (film, television, games). These are all highly competitive global industries. Keeping even a few months ahead of the competition, in terms of innovation and knowledge, is critical to survival, as is the quality of the product and service. Continuous learning is an essential element of a successful work force.

Knowledge Workers

We are now in the era of knowledge workers. By the beginning of the next century, three quarters of the jobs in the U.S. economy will involve creating and processing knowledge. Knowledge workers have already discovered that continual learning is not only a prerequisite of employment but is a major form of work. Some interesting connectors are emerging between technology and learning, between technology and the workplace.

1. Increasingly, work and learning are becoming the same thing; and technology is the connector between these two. Because the new global economy is based on knowledge work and innovation, there is a convergence between work and learning. While you perform knowledge work, you learn. And you must learn minute by minute to perform knowledge work effectively.

2. In the new economy, the learning component of work becomes huge. It includes everything from a software developer creating a new multimedia application, to the manager responsible for corporate planning in a bank, to the consultant assessing a client's markets, to the entrepreneur starting up a new business, or a teaching assistant in a community college (Stewart).

3. Learning is becoming a lifelong challenge as well as a lifelong process. Most knowledge has a shelf life of 3 years or fewer, and knowledge continues to double every 18 months!

Learning Technology Required in the Workplace

A key and rapidly growing resource employed to meet these learning and knowledge needs in the workplace is learning technology. It is anticipated that the worldwide technology training and education market will surge to over $30 billion by the year 2000. Today over 50% of all learning is done via multimedia-based training, while another 35% is done through self-paced video and on-the-job training.

According to a recent survey conducted by Georgia Tech, 34% of organizations are already using the intranet for training. Annual expenditures for software of the intranet has passed $4 billion. Sales within the field of knowledge management could reach $100 billion by the year 2000, says Stan Lepeak, an analyst with Meta Group, Inc.

Organizations have also become more and more "informated," that is, able to immediately acquire information that can be used to get a job done, generate new information as a by-product, and develop new information. An example of informating is a grocery store's use of data scanned from the checkout counter.

The use of telecommunications in training applications will surely increase. Networked computers are increasing communications around the world and global learning organizations use extensive electronic mail networks. Electronic classrooms are also available, allowing ongoing communication between trainees and resource persons in distant locations. In sales training programs, trainees can perform role-playing within a digitally created situation.

Corporate learning in the year 2000 will capitalize on technologies that include computers, multimedia (i.e., audio, animation, and graphics), interactive video, distance learning (providing one-way video and two-way audio communication between an educator and a learner). Corporations will focus on creating learning programs that offer the following dimensions:

- *Modular* — programs that address a single skill rather than a course addressing multiple skills
- *Multisensory* — stimulating sight, sound, and touch in a variety of innovative ways

- *Portable* — moving easily from home to office
- *Transferable* — moving across languages and cultures
- *Interruptible* — having the ability to stop and start easily

Future Challenges

The importance of integrating technology, learning, and organizational life has become increasingly obvious to corporate leaders around the world. According to a recent survey of human resource executives, a number of concerns were expressed:

1. Keeping pace with the rate of change of workplace technologies. Considering the time it takes to develop and implement new technologies and the rapid rate at which they're introduced, these technologies are no longer state-of-the-art by the time they become fully operational (Bassi et al.).
2. Assessing the effectiveness of new learning technologies
3. Knowing when and where to apply new learning technologies
4. Integrating existing technologies with new learning technologies
5. Getting top management to buy-in on using learning technologies

In this book we will explore the capacity of technology for strengthening workplace performance and building corporate success. We will also discuss how the utilization of technology can enhance the speed and quality of learning as well as provide the power to manage knowledge. First, however, we will examine in the next three chapters the myriad ways in which technology specifically affects the workplace, the organization, and the worker.

References

American Society for Training and Development. (November-December, 1997) HRD executives forecast tremendous growth of learning technologies. *National Report on Human Resources.*

Bassi, L., Cheney, S., and Van Buren, M. (November, 1997) Training industry trends 1997. *Training and Development.*

Bates, A.W. (1995) *Technology, Open Learning and Distance Education.* London: Routledge.

Chandrasekaran, R. (November 30, 1997) A seller's market for tech workers. *Washington Post.*

Gross, N. (June 23, 1997) Future of technology. *Business Week.*

Mossberg, W. (September 4, 1997) Personal technology. *Wall Street Journal.*

Stevens, L. (October, 1977) The Intranet: your newest training tool? *Productivity Digest.*

Stewart, T. (1997) *Intellectual Capital: The New Wealth of Organizations.* New York: Doubleday.

Tapscott, D. (1995) *The Digital Economy,* New York: McGraw-Hill.

Toffler, A. (1990) *Power Shift,* New York: Bantam Books.

U.S. Department of Commerce. (1997) *Information Technology Worker Shortage.* Washington, D.C.

Weaver, P. (October 6, 1997) Telecommunications for the new millennium. *Business Week.*

Zuboff, S. (1988) *In the Age of the Smart Machine: The Future of Work and Power.* New York: Basic Books.

2 | 14 Ways Technology Has Transformed the Workplace

echnology has dramatically changed the workplace. Workers no longer need to work in an office. Corporations can collaborate and compete with one another at the same time. Customers can provide supervision as well as dictate services. Fellow employees are able to work closely with each other while never having met one another. Companies have temporary CEOs and part-time strategic planners. Corporate headquarters staff may consist of less than 1% of the company's work force — if there is a headquarters. The GNP is more bytes of information and cyberspace and less manufacturing and products. Organizations are not even "organized." Chaos reigns, the same chaos that is being used as the synergy to propel and energize organizations. This technologically-triggered transformation of business and work has only begun! Let's look in more detail at 14 specific ways in which the workplace and the worker have been re-created because of technology.

1. Technology changes the way work is done, whether it be production, coordination, or management work.

Production work (how to do it) is affected a) by physical supports such as robotics, process control instrumentation, and intelligent sensors, b) by information production such as data processing, and c) by knowledge resources such as CAD/CAM tools.

The key assets of high-value enterprises are not tangible things, but the skills involved in linking solutions to particular needs, and the reputations that come from being successful in the past. Webber notes that the expected distinction between manufacturing and services is becoming less and less real. The real impact of the information economy is to create increasing similarity in the world of work.

Coordination work means that distance and time (time zones) can be shrunk to zero. The organization's memory (command database) can be maintained over time, contributed to from all parts of the organization, and made available to a wide variety of authorized users.

The greatest challenge for a manager in today's environment is to create an organization that can redistribute its knowledge. Intellectual capital, as Stewart notes, is useless unless it moves. By finding ways to make knowledge move, an organization can create a value network, not just a value chain.

Management work is more flexible because information technology can better sense changes in the external environment and stay in close touch with the organization members' ideas and reactions to the environment. Relevant, timely information can be a crucial input for the organization's direction-setting process.

Advances in technology are providing managers with faster transmission of data and expanded storage capacity as well as clearer, more complex links among users and greater computer power. Such innovation will permit greater control of more decentralized organizations, while permitting the information flow needed to give local managers substantive decision-making authority.

Technology also allows more control in two key aspects:

- Measurement (measuring the organization's performance along whatever set of critical success factors has been defined as relevant)
- Interpretation (interpreting such measures against the plan and determining what actions to take)

2. Technology enables a fuller integration of business functions.

Through the power of technology, integration of business functions is possible and assisted at every level within the organization and between organizations. This can be done in four directions:

a. Within the value chain (for example, Xerox connects design, engineering, and manufacturing personnel within its system of local area networks and creates a team focusing on one product. Such teams can finish tasks in a shorter time and with greater creativity and higher morale. With information technology, no part of an organization, in principle, need be excluded from the team concept.)

b.   End-to-end links of value chains (These are links between organizations through just-in-time and electronic data interchange.)

c.   Value chain substitution (These are substitutions via subcontract or alliance.)

d.   Electronic markets (Electronic markets are the most highly developed form of electronic integration, so that travel agents, for example, can electronically reserve seats on all the major carriers and can look around for the best price at which to complete the transaction.)

These four levels of electronic integration have, to varying degrees, the net effect of removing buffers. They also leverage organizational and individual expertise, thereby enhancing individual, team, and organizational learning.

3. Technology creates the possibility of truly global companies.

For globalization to become a reality, companies needed to have the capacity to centralize and decentralize decision-making, so as to be able to connect key personnel around the world synchronously. They had to have a global reach and yet allow implementation at the local level according to local needs. Technology has now made these former impossibilities possible.

Technology allows for what James Moore calls "co-evolution — the notion that by working with direct competitors, customers, and suppliers on a worldwide basis, a company can create new businesses, markets, and industries. Moore urges companies to view themselves as part of a business ecosystem. The new paradigm requires thinking in terms of whole systems. Seeing your business as part of a wider environment, viewing business opportunities not simply from the perspective of a solo player, but as one player among many, each co-evolving with the others — that's sharply different from the conventional idea of competition, in which companies work only with their own resources and do not extend themselves using the capabilities of others. In the global marketplace, technology allows a company to make use of the other players — for capacity, innovation, and capital.

Technology adds considerable capability to the functions of scanning and environmental monitoring. This effective scanning of the business environment to understand what is changing is necessary for an organization to proactively manage its way through a global environment made so turbulent by technological changes. Global communications technology also enables knowledge workers to expedite contact with other knowledge workers, since it is through dialogue and interaction with other knowledge workers that they can refine and improve their ideas, and thereby assist their organizations to an even greater extent.

4. Technology forces basic changes in organizational structure.

Technology demands the re-creation and redefinition of the organizations as we know them. It permits the redistribution of power, function, and control to wherever they are most effective, and according to the mission, objectives, and culture of the organization. Because of technology, corporations can become cluster organizations or *adhocracies*, groups of geographically dispersed people — typically working at home — who come together electronically for a particular project and then disband, having completed their work.

The technology-driven networks and databases will replace the multitiered hierarchy with a wide breadth and depth of knowledge that is the sum of employees' collective experience. The new *organizational architecture* will evolve around autonomous work teams and strategic alliances.

An organization like Digital Equipment Corporation is now able to have all its engineers on the same network. An engineer can share information, ask for help, or work on a project with anyone else in the network. In this way, information technology increases the rate at which information moves and decisions are made.

As more companies realize that the key resources of business are not capital, personnel, or facilities, but rather knowledge, information, and ideas, many new ways of viewing the organization will begin to emerge. Everywhere companies are restructuring, creating integrated organizations, global networks, and leaner corporate centers. Organizations are becoming more fluid, ever shifting in size, shape, and arrangements.

5. Technology enables organizations to transform from bureaucratic to network and federated ways of operating and thinking.

Technology allows for new strategic opportunities for organizations to reassess their missions and operations. It enables organizations to *automate* (which lessens the cost of production), to *informate* (which provides information that can be used to get a job done, generates new information as a by-product, and develops new information), and to *transform*. Morton calls this a stage characterized "by leadership, vision, and a sustained process of organization empowerment."

Technology enables the organization to "stretch," to democratize the strategy-creation process, to tap the imagination of hundreds, if not thousands, of new voices in the strategy process. Technology allows managers to lead their organizations through a complete transformation process, a process that creates the organizational shifts shown in Table 2.1.

Table 2.1   Organizational Shifts Generated by Technology

Dimension	Bureaucratic	Network
Critical tasks	Physical	Mental
Relationships	Hierarchical	Peer-to-peer
Levels	Many	Few
Structures	Functional	Multidisciplinary teams
Boundaries	Fixed	Permeable
Competitive thrust	Vertical integration	Outsourcing and alliances
Management style	Autocratic	Participative
Culture	Compliance and tradition	Commitment and results
People	Homogeneous	Diverse
Strategic focus	Efficiency	Innovation

The integration of technology into the workplace has led to the path of "federalism" as the way to manage increasingly complex organizations in the rapidly changing environment. Charles Handy sees the popularity and success of this way of structuring and sizing organizations as due to the fact that federalism is an effective way of dealing with six paradoxes: 1) power and control, 2) being both big and small at the same time, 3) being autonomous but within bounds, 4) encouraging variety but within a shared purpose, 5) individuality but also partnership, and 6) global and yet local. The structural aspects of federalism have additional benefits:

- Its autonomy releases energy.
- It allows people to be well-informed.
- Its units are bound together by trust and common goals and not by forced control.

- Power is delegated to the lowest possible point in the organization. (A good example of this is Motorola employees who were told by the former Chairman Robert Galvin that they all have the authority of the Chairman when they are with customers.)
- The decentralized structure and interdependence spreads power around and thereby avoids the risks of a central bureaucracy.
- It is very flexible and can never be static.
- Authority must be earned from those over whom it is exercised.
- People have the right and duty to be responsible and recognized for their work.
- Organizations are much flatter (little hierarchy) without losing efficiency.

6. Technology requires new skills and competencies on the part of all workers.

Technology has increased the rapid change of work force skills needed. Over 50% of workers now use computers in their jobs. According to a 1996 ASTD survey, 73% of employers said that computer skills were essential for employment. Simply put, more sophisticated machinery and work processes require more sophisticated workers.

More than half of the new jobs created between 1984 and 2005 will require some education beyond high school. The most significant development is that many of the new jobs will require a much higher level of skill than the jobs they are replacing, especially in manufacturing and resource-based industries. People will retain existing jobs only if they are retrained to higher standards.

As we move from the age of manufacturing to an era of mentofacturing, i.e., where the production is more with the mind (mento) than with the hands (manu), employees are moving from needing repetitive skills to knowing how to deal with surprises and exceptions, from depending on memory and facts to being spontaneous and creative, from risk avoidance to risk taking, from focusing on policies and procedures to building collaboration with people.

A U.S. Department of Labor report, *Economic Change and the American Workforce,* notes: "The competitive workplace today — regardless of the product or service — is a high-skill environment designed around technology and people who are technically competent. Assembly line workers must now understand their work as part of a much larger whole." As mid-level jobs disappear, U.S. society is dividing between high earners, "empowered" in the work force because of their high level of skills, and those in survival wage jobs, consigned to unskilled employment or unemployment.

Peter Drucker sees organizations composed more and more of *knowledge workers*. Not only senior executives, but employees at all levels must be highly educated, highly skilled knowledge workers. In the new post-capitalist society, knowledge is not just another resource alongside the traditional factors of production, land, labor and capital. It is the only meaningful resource in today's work force. In an economy based on knowledge, the knowledge worker is the single greatest asset.

Tomorrow's workers, according to Reich, will more and more need to have the three skills essential in driving the emerging businesses, all of which will be enhanced with technological capabilities:

- Problem-identifier skills (required to help customers understand their needs and how those needs can best be met by customized products)
- Problem-solving skills (required to put things together in unique ways)
- Strategic-broker skills (needed to link problem-solvers and problem-identifiers)

Technology will require "higher-order cognitive skills — the ability to analyze problems and find the right resources for solving them, and often with both limited and conflicting information.

The demands of technology and customer satisfaction will continue to recreate and redefine jobs. For example, computer access to financial information means that entry-level workers are no longer tellers, but customer service representatives. They must know the full range of the bank's services, be able to assess customers' financial needs and investment expectations, and find a match between the two. Computer innovation will empower employees low on the organizational ladder. These workers are closest to the point of production and service and thus need to gain access to more data and be entrusted with greater decision-making responsibility.

7. Technology impacts where workers work.

"Welcome to the Officeless Office" proclaims a headline in a recent *Business Week* story. Telecommuting, thanks to digital phone lines, affordable desktop videoconferencing and wide-ranging cellular networks, is out of the experimental stage. By means of local phone companies offering Integrated Services

Digital Network (ISDN) lines that can transmit voice, data, and video simultaneously, telecommuting has become easy and highly productive.

The over 15 million telecommuters in the U.S. represent the fastest-growing portion of workers. The entire 240-member core sales staff at American Express Travel Related Services Co. are telecommuters. Ernst & Young has implemented "hoteling," in which up to 10 people share a single desk in a fully equipped office on an as-needed basis. Employees must reserve space and equipment in advance. Over the past 3 years, the accounting firm has slashed its office space requirements by about 2 million square feet, saving roughly $25 million a year.

In addition to reducing air pollution and cutting down on office space and equipment purchases, telecommuting enables corporations to hire otherwise unavailable key talent. For example, Northern Telecom in Memphis was able to hire someone from Philadelphia who did not want to move to Tennessee.

8. Technology provides more opportunities and power to customers.

Technology not only has provided customers with much more knowledge about the various possibilities and standards relative to products and services, but also allowed them easy access to purchasing via shopping channels on TV or shopping carts on the Internet. Amazon.com is the largest bookstore in the world with over 3 million books available for instant purchase.

Technology permits customers to demand "customized" products and services. Customers now have many choices, and are making ever more challenging demands. They choose the products and services they want based on the best:

a. *Cost* — what is the least expensive and most economical
b. *Quality* — no defects; meeting and exceeding the customer's expectations
c. *Time* — available as quickly as possible
d. *Service* — pleasant, courteous, available, and on products which are reparable or replaceable
e. *Innovation* — new, something not yet envisioned by the customer when produced (e.g., Sony Walkman)
f. *Customization* — tailored to very specific needs

9. Technology allows for the emergence of virtual organizations.

Technology makes it easier to form virtual organizations, a form of restructuring that is rapidly gaining popularity. Virtual organizations are a temporary network of independent companies, suppliers, customers, even rivals linked by information technology to share skills, costs, and access to one another's markets. In its purest form, a company decides to focus on the thing it does best. Then it links with other companies, each bringing to the combination its own special ability. The virtual corporation has neither central office nor organization chart, and no hierarchy or vertical integration. Teams of people in different companies routinely work together. After the business is done, the virtual organization disbands.

A virtual organization mixes and matches what it does best with the best of the other companies. For example, a manufacturer will manufacture, while relying on a product-design outfit to sell the output. Such a best-of-everything organization, with the power of technology, can become a world-class competitor, with the speed, the muscle, and the leading-edge technology to pounce on the briefest of opportunities. Organizational theorists believe that the virtual model could become the most important organizational innovation since the 1920s when Pierre DuPont and Alfred Sloan developed the principle of decentralization to organize giant complex corporations.

10. Technology affects reward systems of workers.

Technology will place greater emphasis on contributions to discrete projects as the measure of an employee's value to the corporation, thereby strengthening the rationale for pay-for-performance and gainsharing schemes. Technology allows us to record and measure in much greater detail each individual's activities and the myriad impact of those efforts.

11. Technology transfers knowledge faster and more efficiently between workers and throughout the organization.

Technology provides for quicker and better transfer of knowledge throughout the organization for many reasons. First, technology can improve the ability of people to communicate with one another because it blurs the boundaries of the company and increases the range of possible relationships beyond hierarchies. Second, technology makes it easier for people to communicate directly with one another across time and space through such media as electronic mail and videoconferencing. Third, it reduces the number of management levels needed in the hierarchy, while at the same time providing an enhanced potential for span of control. Empowered with information, the front line workers can become much more autonomous. Finally, technology contributes to flexibility with mobile workstations, relational databases, and the storage of knowledge in open databases rather than in the minds of individuals. Database, texts, articles, reports, manuals, and directories can be held for quick and easy access by all workers.

Personal Area Network

IBM has developed a "personal area network" (PAN) that lets two business persons exchange a calling card's worth of personal information simply by shaking hands. Both must carry card-sized transmitters and receivers. Their handshake then completes an electric circuit, and each person's data are transferred to the other's laptop computer.

---

12. Technology affects how training is designed and delivered.

Training, because of technology, can be done anywhere, anytime, for anyone. Technology significantly impacts the way training programs are designed, delivered, and evaluated. Needs analysis may be done at a distance and by computer. Objectives need to be established for a wide diversity of people located throughout the world, yet customized for each individual. Design becomes more important and costly since the number of people receiving a program can exceed the tens of thousands. Development of training materials requires much more sophistication and familiarity with the strengths and limitations of each of the learning technologies.

Instructors need to be effective and stimulating even though they may not be physically in front of the learners. Since technology allows delivery to be done asynchronously as well as synchronously, delivery and interactions need to adjusted. Finally, technology provides administrators with the opportunity to continuously monitor and assess the progress and capabilities of the participants through a variety of software. In Part II we will discuss how the various learning technologies can be used for the design and delivery of training programs.

13. Technology affects how knowledge is managed.

The rapid increase in the volume and quality of knowledge required by workers to do their work is literally impossible to accomplish without technological assistance. More knowledge is accessible and storable, but a growing competence is the ability to distinguish between "garbage data" and "valuable knowledge."

Technology allows organizations to collect, store, analyze, transfer, and apply knowledge with magnificent efficiency and wider availability to workers. In Part III we will examine in depth how technology can manage knowledge and provide several case studies to demonstrate ways in which companies can use technology for knowledge management.

14. Technology affects how organizations learn.

Becoming a learning organization is now recognized as absolutely essential for an organization to compete and to succeed in the global arena. Becoming a learning organization, of course, requires more than technology. However, if the other subsystems of a learning organization are present (i.e., empowered and enabled people; corporate learning culture, vision, and strategies; an info-structure that creates and supports learning; learning in all parts of the business chain; various types of learning at all levels of the organization, etc.), technology can become the turbocharger that empowers an engine.

Technology truly is the catalyst and enhancer of building the learning organization since it contributes some unique features for facilitating organizational learning. First, it can provide fast, efficient communications, bridging space and time. Second, technological systems have the capability of creating a stored history of problem-solution exchange experiences for all members. Third, technology-aided systems provide a mechanism where multiple members dynamically share solutions and update their problem–solution experiences (Goodman). In Chapter 3, we will describe how technology helps to create and support the learning organization.

The New Workplace

These 14 impacts have transformed the workplace into a totally new environment — one where new kinds of people work in new kinds of locations, where new resources are used for new purposes. It is a workplace that can better tap the full capabilities and energies of the worker, which gives each person power and excitement and adds meaning to work. Technology should not be seen as something to replace workers in the workplace, but rather to enhance them and enable them to better fulfill their full potentials as human beings. The most promising and productive workplace in which to be a worker is the learning organization, which we will explore in Chapter 3.

References

Drucker, P. (September-October, 1992) The new society of organizations. *Harvard Business Review.*

Goodman, P. and Darr, E. (1996) Computer-aided systems for organizational learning, in *Trends in Organizational Behavior,* Cooper C. and Rousseau, D. (Eds.), New York: John Wiley & Sons.

Handy, C. (1989) *The Age of Unreason.* London: Basic Books.

Morton, M. (1991) *The Corporation of the 1990s.* New York: Oxford University Press.

Reich, R. (1991) *The Wealth of Nations.* New York: Random House.

Stewart. T. (1997) *Managing Intellectual Assets.* New York: McGraw-Hill.

Webber, A. (January-February, 1993) What's so new about the new economy, *Harvard Business Review,* pp. 24-42.

Welcome to Officeless Offices. (June 26, 1995) *Business Week,* p. 104.

3 Technology and the Learning Organization

The immense changes in the social and economic environments caused by technology and globalization have forced organizations worldwide to make overwhelming changes relative to their purpose, strategies, and even structures in order to adapt, survive, and succeed in the 21st century. Organizations must not only become wired, retooled, and networked, they also need to change both the extrinsic elements of a company — products, activities, or structures — and their basic, intrinsic way of operating — values, mindset, even their purpose.

Harrison Owen predicted this significant shift in *Riding the Tiger: Doing Business in a Transforming World*, when he wrote, "There was a time when the prime business of business was to make a profit and a product. There is now a prior, prime business, which is to become an effective learning organization. Not that profit and product are no longer important, but without continual learning, profits and products will no longer be possible. Hence, the strange thought: the business of business is learning — and all else will follow." (Owen, p. 1) Technology must serve as the catalyst and supporting beam in building the learning organization, because only technology can give organizations the required speed and power to deal with change

Put very bluntly, organizations must learn faster in order to adapt to rapid environmental changes or they simply will die. As in any transitional period, there presently exists side-by-side the dominant, dying species (i.e., non-learning organization) and the emerging, more adaptive species (i.e., learning organization). Within the next 10 years, only learning organizations will have the adaptive capacities to survive. Companies that do not become learning organizations will soon go the way of the dinosaur; they will die because they were unable to adjust to the changing atmosphere around them.

Why Organizational Learning Is so Critical

The demands put on organizations in today's world require learning to be delivered faster, cheaper, and more effectively to a fluid workplace, and to a mobile work force more profoundly affected by daily changes in the marketplace than ever before. Knowledge is doubling every 2 to 3 years. Astounding breakthroughs in new and advanced technologies become more frequent. Global competition forces companies to face the knowledge resources of the world's best companies. Change is more rapid and ever more intense!

As Reginald Revans, a pioneer in organizational learning notes, "Learning inside must be equal to or greater than change outside the organization, or the organization declines and dies."

Only through company-wide, systems-wide learning can organizations survive. The Rover Group recognized this fact in the early 1990s when its chairman, Sir Graham Day, proclaimed that "if our company seeks to survive and prosper, learning is essential." He further stated that "the prospect that organizational learning offers us is one of managing change by allowing for quantum leaps. Continuous improvement means that every quantum leap becomes an opportunity to learn and therefore prepare for the next quantum leap. By learning faster than our competitors the time span between leaps reduces and progress accelerates."

To obtain and sustain competitive advantage in this new environment, organizations will have to learn better and faster from their successes and failures. They will need to continuously transform themselves into a learning organization, to become places where groups and individuals continuously engage in new learning processes.

Shoshanna Zuboff, in her classic *In the Age of the Smart Machine*, writes how today's organization may indeed have little choice but to become a "learning institution, since one of its principal purposes will have to be the expansion of knowledge — not knowledge for its own sake (as in academic pursuit), but knowledge that comes to reside at the core of what it means to be productive. Learning is no longer a separate activity that occurs either before one enters the workplace or in remote classroom settings. Nor is it an activity preserved for a managerial group. The behaviors that define learning and the behaviors that define being productive are one and the same. Learning is the heart of productive activity. To put it simply, learning is the new form of labor." (Zuboff, p. 395)

What's New About Learning Organizations

There are seven key paradigm shifts that make a learning organization different from the traditional organization.

Traditional Focus	*Learning Organization Focus*
Productivity	Quality of performance
Workplace	Learning environment
Predictability	Systems and patterns
Training	Learning
Worker	Continuous learner
Supervisor/manager	Coach and learner
Engagement/activity	Learning opportunity

As a result of these paradigm shifts, there is a whole new mindset and way of "perceiving" organizations and the interplay between "work" and "learning." Learning must take place as an ongoing by-product of people doing their work — in contrast to the traditional approach of acquiring knowledge before performing a particular task or job (Marquardt).

The learning in organizational settings therefore represents a "new form of learning" in the following ways:

1. It is performance-based and clearly tied to business objectives.
2. Importance is placed on learning processes and skills, especially learning how to learn.
3. The ability to define learning needs is as important as the answers.
4. Organization-wide opportunities are created to develop knowledge, skills, and attitudes.
5. Learning is part of work and a part of everybody's job description.

The need for individuals and organizations to acquire more and more knowledge will continue unabated, but what people and organizations *know* takes second place to what and how quickly they can *learn*. Learning skills will be much more important than data. Penetrating questions will be much more important than good answers.

A learning organization has the powerful capacity to collect, store, and transfer knowledge and thereby continuously transform itself for corporate success. It empowers people within and outside the company to learn as they work. A most critical component is the utilization of technology to optimize both learning and productivity.

There are a number of important dimensions and characteristics of a learning organization:

- Learning is accomplished by the organizational system as a whole, almost as if the organization were a single brain.
- Organizational members recognize the critical importance of ongoing organization-wide learning for the organization's current as well as future success.
- Learning is a continuous, strategically used process — integrated with and running parallel to work.
- There is a focus on creativity and generative learning.
- Systems thinking is fundamental.
- People have continuous access to information and data resources that are important to the company's success.
- A corporate climate exists that encourages, rewards, and accelerates individual and group learning.
- Workers network in an innovative, community-like manner inside and outside the organization.
- Change is embraced, and unexpected surprises and even failures are viewed as opportunities to learn.
- It is agile and flexible.
- Everyone is driven by a desire for quality and continuous improvement.
- Activities are characterized by aspiration, reflection, and conceptualization.
- There are well-developed core competencies that serve as a taking-off point for new products and services.
- It possesses the ability to continuously adapt, renew, and "revitalize" itself in response to the changing environment (Marquardt and Reynolds; Pedler et al.).

Technology Vital for Learning Organizations

There is no way an organization can adapt to the rapid change in the environment without sound and solid use of technology. Therefore, learning organizations can only be created and succeed with the intelligent application

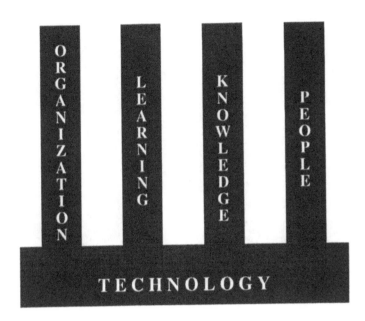

Figure 3.1   Technology as Foundation to a Learning Organization

of technology. Technology is necessary to create the appropriate learning organization "info-structure" to enhance the speed and quality of needed learning as well as to adequately manage the information and knowledge of the organization. In Figure 3.1, we can see how technology serves as the foundation for building the learning organization.

Technology not only serves as a foundation, but also as the key integrating system, including the supporting networks and tools that allow access to and exchange of information and learning. It includes technical processes, systems, and structures for collaboration, coaching, coordination, and other knowledge skills. It encompasses electronic tools and advanced methods for learning, such as computer conferencing, Internets and intranets, electronic publishing, and multimedia learning, all of which work to create knowledge freeways. These technological tools will be explored in depth in Part II (learning technology tools) and in Part III (knowledge management technology tools). At this time, let's look at the four subsystems of the learning organization supported by technology, namely: 1) organization, 2) people, 3) learning, and 4) knowledge.

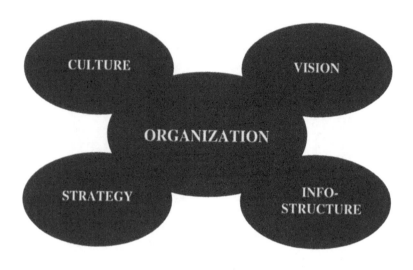

Figure 3.2   Dimensions of Organization

Organization and the Info-Structure

The organization is the setting and body in which the learning occurs. As Peter Drucker notes, "Like a blast furnace that converts iron and coke into steel, the organization concentrates, processes, and reifies the knowledge worker." There are four key dimensions of this subsystem of the learning organization: culture, vision, strategy, and info-structure. These are shown in Figure 3.2.

a. *Culture* refers to the values, beliefs, practices, rituals, and customs of an organization. It helps to shape behavior and to fashion perceptions. In a learning organization, the corporate culture is one in which learning is recognized as absolutely critical for business success, where learning has become a habit and an integrated part of all organizational functions. This rich, adaptable culture creates integrated relationships and enhances learning by encouraging values such as teamwork, self-management, empowerment, and sharing. It is the opposite of a closed, rigid, bureaucratic architecture.

   Probably no company has put as much effort and been as successful in creating a learning culture as Royal Bank of Canada. At Royal Bank, learning is a committed three-way partnership among the employee, the manager, and the human resources department. Learning opportunities are available at all times inside and outside of the Bank. As James Gannon, Vice President of Human Resources Planning and Development notes, "Learning has to become a way of life rather than a once-in-a-while type of event."

b. *Vision* captures a company's hopes, goals, and direction for the future. It is the image that the organization transmits inside and outside the organization. In a learning organization it depicts and portrays the desired future picture of the company, a picture in which learning and learners create the company's continuously new and improving products and services.

Singapore Airlines, perhaps the world's most successful airline, exemplifies how a clear vision can create and transfer powerful learning throughout the organization. Corporate philosophy and documents are filled with statements emphasizing the importance of learning for present and ongoing corporate success. The company spends over $100 million a year in employee-learning programs and has won many awards for its service superiority.

c. *Strategy* relates to the action plans, methodologies, tactics, and steps employed to reach a company's vision and goals. In a learning organization, these are strategies that optimize the learning acquired, transferred, and utilized in all company actions and operations.

In traditional hierarchical organizations with multiple layers of management, accountability, and bureaucracy, information flow is vertical. Host-based islands of technology correspond to the old structures. In a learning organization, the corporate pyramid is replaced with internetworked teams using groupware technology. The focus shifts from the individual who is accountable to the manager to teams that function as service units to clients and to other teams both internal and external.

Technology enables designers to collaborate around workstations to engineer a new design concurrently rather than in serial fashion. Many functions can happen simultaneously; individuals are not waiting for someone else to complete a task before they add their input.

d. *Info-structure* includes the departments, levels, and configurations of the company. A learning organization is a streamlined, flat, boundaryless structure that maximizes contact, information flow, local responsibility, and collaboration within and outside the organization. The structure in a learning organization is much different, thanks to the power of technology. The network and database replace the depth of knowledge a multitiered hierarchy offers with the breadth of knowledge that is the sum of employees' collective experience.

Technology enables an organization to move beyond the old hierarchy because layers of management are not required when information is instantly available electronically. An info-structure can enable the enterprise to function as a cohesive organization by providing corporate-wide information for decision making and new competitive enterprise applications that transcend autonomous business units or teams.

Hewlett-Packard, once a "lumbering dinosaur" in terms of structure and innovation, has become, according to *Business Week,* "gazelle-like" with speed of learning a top priority. Learning teams now rethink every process from product development to distribution. Asea Brown Boveri (ABB) and General Electric have taken similar steps to restructure themselves into learning organizations.

Empowered and Enabled People

A learning organization involves people throughout the entire business chain of the company; i.e., employees, managers/leaders, customers, business partners (suppliers, vendors, and subcontractors), and the community itself (Figure 3.3). Each of these groups is valuable to the learning organization, and all are empowered and enabled to learn with the help of technology.

a. *Employees* as learners are expected to and are helped to learn continuously, and are rewarded for their efforts.

Honda is an exemplary company in empowering its people. Honda does not just talk empowerment; it permits people to set out and create the new cars. Robert Simcox, a plant manager, says that Honda people are learning together because they have been "given the power to use their own creativity and imagination."

b. *Managers/Leaders* as learners carry out coaching, mentoring, and modeling roles with the primary responsibility for generating and enhancing learning opportunities for people around them.

Figure 3.3 People in the Learning Organization

c. *Customers* as learners participate in identifying needs, receiving training, and being linked to the learning of the organization. They are a key source in predicting and preparing organizations for future changes. Companies like Whirlpool, 3-M, and Goodyear actively solicit and connect their customers to the learning cycle.

d. *Suppliers and vendors* as learners can receive and contribute to instructional programs.

By linking their computer systems with their most important suppliers, stores such as Wal-mart and Target have included customers in the learning equation. Technology enables their suppliers to forecast demand for their products as well as to help the retailers strengthen their supply networks, reduce inventory, and improve product availability on the shelves.

e. *Community groups* as learners include social, educational, and economic agencies that can share in providing and receiving learning.

Rover devotes significant resources to provide learning opportunities for customers, dealers, suppliers, and the community. Recently, Rover launched a quality management program for customer service initiative. The program provides a structured career path, via a "learning and competence accreditation ladder." In its first, year, over 2000 dealer staff enrolled in what has been described as a "remarkable confirmation of the continuous learning ethos" at Rover.

Figure 3.4   Learning Dynamics

Learning Dynamics

The core of the learning organization is learning itself. In Part II of this book, we will discuss the various technologies that can assist in workplace learning. At this time, we will describe only the a) levels of learning, b) types of learning crucial for organizational learning, and c) critical organizational learning skills (Figure 3.4).

Levels of Learning

There are three levels of learning that are supported in learning organizations:

Individual learning refers to the skills, insights, knowledge, attitudes and values acquired by a person through self-study, technology-based instruction, and observation.

Technology has significantly enhanced the effectiveness and efficiency of learning at the individual level. Personal computing has become multimedia computing with a rich and natural power of audio, image, and video integrated into compound digital documents and humanized styles of computing. Most new computers are powerful enough to handle new forms of input beyond the keyboard, such as microphones, scanners and cameras. Individuals can do much more in less time. For example, with multimedia-based learning tools one can learn something complex in about half the time, yet retain that knowledge two to three times longer.

Group or team learning alludes to the increase in knowledge, skills, and competence accomplished by and within groups. Through team learning, business teams can generate faster responses to changes in the business environment and increasing customer demands (Watkins and Marsick). The killer application to launch workgroup learning and computing was the software product known as Lotus Notes. Notes enabled workgroups to form and be immensely effective.

Organization learning represents the enhanced intellectual and productive capability gained through corporate-wide commitment and opportunity to continuous improvement (Dixon; Redding). It differs from individual and group/team learning in two basic respects. First, organizational learning occurs through the shared insights, knowledge, and mental models of members of the organization. Second, organizational learning builds on past knowledge and experience — that is, on organizational memory that depends on institutional mechanisms (e.g., policies, strategies, and explicit models) used to retain knowledge (Marquardt).

Types of Learning

There are several types or ways of learning that are significant and add value to the learning organization. Although each type is distinctive, there is often overlap and complementarity. Therefore, a particular learning occurrence may be classified as being of more than one type. For example, action learning may be classified as also adaptive or anticipatory.

Adaptive, anticipatory, and generative learning is learning from experience and reflection. Anticipatory learning is the process of acquiring knowledge from expecting the future (a vision-action-reflection approach), whereas generative learning is the learning that is created from reflection, analysis, or creativity.

Single-loop, double loop and deutero learning are differentiated by the degree of reflection placed on action that has occurred in the organization.

Action learning involves reflecting on real problems using the formula of L(learning) = P (existing or programmed knowledge) + Q (questioning insight). Action learning involves a learning team, questioning processes, reflection on real organizational problems, a commitment to action and learning, and guided facilitation.

Skills of Organizational Learning

Peter Senge identified several key skills (or disciplines) to initiate and maximize organizational learning, namely:

Systems thinking, which represents a conceptual framework one uses to make full patterns clearer, and to help one see how to change them effectively.

Mental models are the deeply ingrained assumptions that influence how we understand the world and how we take action. For example, our mental model or image of "learning" or "work" or "patriotism" impacts how we relate to and act in situations where those concepts are present and operating.

Personal mastery indicates the high level of proficiency in a subject or skill area. It requires a commitment to lifelong learning so as to develop an expertise or special, enjoyed proficiency in whatever one does in the organization.

Team learning focuses on the process of aligning and developing the capacity of a team to create the learning and results its members truly desire.

Shared vision involves the skill of unearthing shared "pictures" of the future that foster genuine commitment and enrollment rather than compliance.

Dialogue denotes a high-level of listening and communication between people. It requires the free and creative exploration of subtle issues, a deep listening to one another, and suspending one's own views. The discipline of dialogue involves learning how to recognize the patterns of interaction in teams that promote or undermine learning. For example, the patterns of defensiveness are often deeply ingrained in how a group of people or an organization operates. If unrecognized or avoided, they undermine learning. If recognized and faced creatively, they can actually accelerate learning. Dialogue is the critical medium for connecting, inventing, and coordinating learning and action in the workplace.

Perhaps no organization has incorporated the various aspects of the learning subsystem as well as Arthur Andersen Worldwide. Learning better and faster has become the firm's key focus. Staff development at Andersen centers on the learner who, as a decision maker, chooses from among various available tools and resources to learn what he or she needs for success.

According to Joel Montgomery, senior education specialist at Andersen's Center for Professional Education, learners are now "much more active in the learning process, and are jointly responsible for their learning. Learners are asked to use what they have learned rather than repeating or identifying what they have been exposed to." Andersen encourages and initiates all types and levels of learning. The company designs its learning programs in a way that stimulates the learners to engage in activities that allow them to focus their learning on what they know they need. In the process, they are given the skills and tools to reflect on what they are doing, to evaluate it according to some standard, and to give and receive feedback about what they are doing and learning. This ensures a greater depth of learning.

Knowledge Management

The knowledge subsystem of a learning organization refers to the management of acquired and generated knowledge of the organization. It includes the acquisition, creation, storage, transfer, and utilization of knowledge (Figure 3.5). We will briefly describe these components. In Part III of this book, we will explore in greater depth the role technology plays in managing knowledge, as well as discuss specific technologies for managing knowledge such as electronic performance support systems and the networks (Internet, intranet, worldwide web, and LANs/WANs).

Acquisition is the collection of existing data and information from within and outside the organization via benchmarking, conferences, environmental scans, use of Internet, and staff suggestions. Everyone adds to the knowledge base, and thereby increases the corporate memory and database. It is important that the organization provide an environment that encourages the collection of

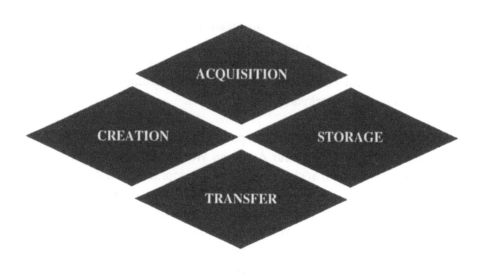

Figure 3.5 Knowledge Management in the Learning Organization

knowledge. Rover, Andersen, Xerox, and Sony regularly benchmark and systematically participate in information exchange programs.

Creation refers to new knowledge that is created within the organization through problem-solving, time set aside for working on innovative programs, demonstration projects, and conversion of implicit knowledge to explicit knowledge (Nonaka and Takeuchi).

Storage is the coding and preserving of the organization's valued knowledge for easy access by any staff member, at any time, and from anywhere.

Transfer and utilization refers to the mechanical, electronic, and interpersonal movement of information and knowledge, both intentionally and unintentionally, throughout the organization as well as its application and use by members of the organization. Information is unrestricted and everyone is encouraged to use company databases.

The knowledge elements of organizational learning are ongoing and interactive instead of sequential and independent. The collection and distribution of information occurs through multiple channels, each having different time frames. An example is an on-line newsletter that systematically gathers, organizes, and disseminates the collective knowledge of the organization's members.

Knowledge transfer is a part of everyone's job and is considered part of the personnel evaluation process at McKinsey & Company. National Semiconductor holds "sharing rallies," in which the best programs of each plant are presented at local, and then regional and international, levels.

Technology Facilitates and Enhances Organizational Learning

The organization that makes learning its core value can rapidly leverage its new knowledge into new products, new marketing strategies, and new ways of doing business. Learning organizations will become the only place where global success is possible, where quality is more assured, and where energetic and talented people want to be.

In the final analysis, what makes a company a learning organization is 1) its capacity to have its people learn faster and better and 2) its ability to effectively manage knowledge. Technology is absolutely necessary to accomplish each of these functions. In the remaining chapters of the book, we will examine in depth how technology provides this support.

References

Dixon, N. (1994) *The Organizational Learning Cycle.* New York: McGraw-Hill.

Marquardt, M. (1996) *Building the Learning Organization.* New York: McGraw-Hill.

Marquardt, M. and Reynolds, A. (1994) *The Global Learning Organization.* Burr Ridge: Irwin Publishing.

Nonaka, I. and Takeuchi, H. (1995) *The Knowledge-Creating Company.* New York: Oxford University Press.

Owen, H. (1991) *Riding the Tiger: Doing Business in a Transforming World.* Potomac: Abbott Publishing.

Pedler, M., Burgoyne, J., and Boydell, T. (1991) *The Learning Company.* London: McGraw-Hill.

Redding, J. (1994) *Strategic Readiness: The Making of the Learning Organization.* San Francisco: Jossey-Bass.

Revans, R. (1980) *Action Learning.* London: Blond & Briggs.

Senge, P. (1990) *The Fifth Discipline.* New York: Doubleday.

Watkins, K. and Marsick, V. (1993) *Sculpting the Learning Organization.* San Francisco: Jossey-Bass.

4 Technology and the Worker

Technology impacts nearly every aspect of the worker in the workplace — what one does, the location of work, whom one works with, what and how the workers learn, the quality of worklife, etc. In this chapter, we will explore five key issues that emerge as the worker interacts with technology:

1. The organizational impact of technology on the worker
2. How workers interact with technology
3. Accommodating special needs
4. Security and ethics issues created by technology
5. Technoculture and technophobia

Impact of Technology on the Worker/Learner

Technology allows us to break many of the old rules of decision-making, labor/management differences, and customer involvement. Why? Because technology allows the following to happen:

1. Information can appear simultaneously in as many places as needed.
2. A generalist can do the work of an expert.
3. Organizations can simultaneously reap the benefits of centralization and decentralization.
4. Decision-making is part of everyone's job.
5. Field personnel can send and receive information wherever they are.
6. Plans can be revised instantaneously.

As noted in Chapter 2, technology significantly affects the skills and knowledge competencies needed by today's workers. It also determines where they work (more and more are working at home or on the road) and the reward systems and policies of the company.

How Workers Interact with Technology

Computers have become ubiquitous in the workplace. It's estimated at least half of all workers use some form of computer-based system in their jobs today — with the percentage growing all the time. When you add the interaction with bank ATMs and electronic cash registers in stores, almost everyone now has frequent contact with technology. And this does not take into account technologies such as fax machines, cellular phones, VCRs/camcorders, or other electronic gismos that most people have in their offices or homes.

This raises a very fundamental question: How do people learn to use all these technologies? We expect every individual to have a certain level of understanding of technology in order to cope in today's society — but where and when is this knowledge and competence to be acquired? In every organization, the issue of technology skills is a basic concern, and a great deal of planning, resources, and assessment is devoted to the subject. Although more and more workers are provided with access to computer systems and expected to use them, there is little systematic analysis of how people learn to use computer-based systems and what methods seem to be most or least effective.

Four Levels of Technology Interaction

While there is tremendous diversity in how people use technology, we can identify four different types or levels of interaction (Figure 4.1). The first involves interaction with computer-based equipment such as electronic cash registers or bar-code scanners, robotic or computer-controlled machine tools, computerized diagnostic devices, and various office machines. At this level, people interact directly with hardware but usually not with software. The level of technology understanding needed is relatively minimal other than a specific set of procedures or steps required to operate the equipment. This level of interaction is typical of service and manufacturing jobs, as well as among the general public.

At the second level is interaction with one or more proprietary application programs. This would include any system for processing transactions (e.g.,

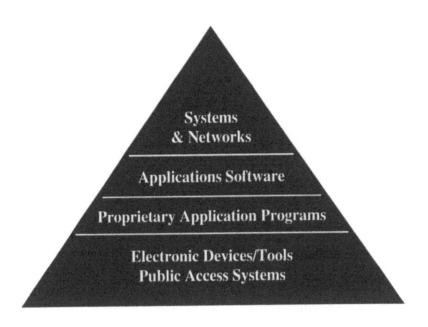

Figure 4.1 Levels of Technology Interaction

customer orders, billing, reservations), completing or checking records in a database (e.g., medical, insurance, law enforcement, inventory), or specialized decision-making (e.g., loan appraisals, budgeting, transportation routing). The information processing activities of almost every organization depend heavily on such proprietary software and millions of clerical jobs revolve around their use. Learning to use such a system typically requires formal training delivered by the organization (internal staff or contractors). The duration and extent of this training can range from a few hours to many weeks depending on its complexity and stability.

The third level involves the use of commercially available application software such as word processing, spreadsheets, databases, project management, or telecommunications programs. While organizations may use these programs in unique or specialized ways (including interfaces to proprietary systems), they come with their own training materials including manuals, on-line helps/tutorials, and workshops. Learning how to use application software usually takes some time, although achieving basic proficiency may only take a few days. The skills and knowledge that individuals acquire when they learn to use these kinds of programs are transferable across organizations and jobs — as well as specific brands of software. Students often learn how to use application programs in school and college and bring that competency with them to their jobs. Most administrative staff, as well as many managers and professionals, are comfortable using a broad selection of applications software.

The fourth level entails a knowledge of computer systems or networks themselves. This is the domain of information or technology specialists who plan, design, procure, install, or maintain hardware and software. They may be programmers, technicians, engineers, systems analysts, project managers, or senior administrators. The range of specific skills or knowledge about technology required in these positions can be very extensive and diverse. It is also likely to be very transient, since the hallmark of technology is the speed at which everything changes. This means that individuals in this category must be perpetually learning — through formal training and their own self-directed efforts. In theory, they must possess all the knowledge and skills of people at the other three levels in order to implement and manage the systems they use. In practice, they often don't, which foreshadows some of the issues and problems we will discuss later in this chapter.

As Figure 4.1 shows, the largest percentage of users is in the first category indicated at the bottom of the pyramid, with the smallest number in category four at the top. The extent of technology skills and technology required increases as you go up in the pyramid and so does the amount and level of computer-related training needed to be a competent worker. The four levels also correspond to increases in salary and career opportunities. Of course, there are plenty of occupations that do not fit into this structure, such as business owners/entrepreneurs, athletes, entertainers, and politicians. But even those who do not use computers themselves usually depend on others who do.

The critical issue as far as productivity is concerned is how should people at each of these four levels be trained? Or more generally, what is the best way to prepare individuals for jobs and careers that involve interaction with technology. In many cases, how well people are able to do their jobs depends on their technology skills. If the cashier in a grocery story is skilled at operating his/her electronic cash register and bar-code scanner, he/she can process orders faster, keep customers happy, and make fewer mistakes. If a claims processing clerk understands how to use a claims processing system well, we have the same type of outcome. If a project administrator is competent in the use of a project management program, resources and budgets will be easily tracked, and projects can be successfully supervised. If a systems analyst designs or develops a new program properly, it will achieve the intended results, be easy to use, and not fail in unexpected ways. Each level of interaction presents very different training challenges and learning needs.

Addressing IT Skills at a National Level: Canada's SHRC and Mentys

Ironically, the job domains most affected by technology change are information technology (IT) workers. Given how quickly new hardware and software appears, it is very difficult for systems engineers, computer technicians, software developers, and programmers to keep up to date in their job skills. IT professionals must be committed to lifelong learning to be successful in this field.

The Canadian government has responded to the challenge of providing its IT work force with up-to-date skills through the creation of the Software Human Resource Council (SHRC), a special government corporation charged with the mission of identifying the IT skills needed and figuring out how to create training programs to address the skill gaps. To do this, SHRC chose 14 industry partners to develop and deliver IT training courses. One such effort is Mentys, an Internet-based CBT system that provides a variety of courses in operating systems, programming, and network architecture. When individuals sign up for Mentys training courses, they first go through the Competence Key, a skills assessment that determines current and needed skills relative to a specific job or career. Based on the results of the assessment, the system recommends specific courses available to meet skill deficiencies.

Mentys lessons are designed so they can be completed in 30-minute increments, making it possible to fit learning into the busy work or personal lives of IT professionals. After completing a lesson, a test indicates what aspects were not understood and points the learner back to the relevant parts of the lesson. The system also features an on-line conferencing capability so participants can discuss topics/issues with content experts, as well as other participants.

To date, hundreds of people have tried Mentys courses and appear to be satisfied. However, the system developers (Global Knowledge Network) readily acknowledge that CBT is not for everyone, although it is well suited to the nature of IT professionals, who tend to be highly motivated learners capable of independent study. They also acknowledge that a large-scale CBT system like this requires constant upgrading because the content changes constantly as do the skills required for IT jobs/careers. This is why an organization such as the SHRC is necessary to provide the funding and coordination for efforts such as Mentys.

L. Eline, Bridging the Gap in Canada's IT Skills
Technical & Skills Training, July 1997.

Interaction with Others Via Technology

So far we have been discussing interaction with technology, but how people use technology to interact with each other is an equally important issue (Galegher et al.; Rosenberg). Through the use of electronic mail, teleconferencing, and groupware programs, individuals in an organization can interact in a cost-effective manner and increase their productivity and that of the organization. For example, e-mail allows people to disseminate information to individuals, small groups, or the entire organization with no extra effort. This greatly improves the potential for communication and coordination within or across organizations. While in theory the same results can be achieved via regular mail, fax, or telephone calls, there is some extra measure of effort associated with each individual contacted; but this is not true with e-mail (assuming distribution lists have already been created). Furthermore, since e-mail is an asynchronous technology (see detailed discussion in next chapter), it is not necessary for recipients to be available to receive their messages because they are stored until they are read. This is very important in terms of contacting individuals who are very mobile, have variable work schedules, or are spread across time zones.

Similar kinds of benefits result from teleconferencing (see Chapter 8 for a full discussion of this topic). Use of teleconferencing reduces the need for travel and associated expenditures. People can conduct meetings that involve multiple individuals at different locations (even internationally) without all of the headaches involved with travel. However, for teleconferences to be successful, all of the participants have to be comfortable with the technology used and with the idea of interacting through the technology. These skills and understanding are not automatic, and it takes some practice before most people develop them.

Interacting via e-mail takes time to get used to as well. First of all, people must learn to be concise in their writing — long, wandering messages will not be read. Expressing themselves in written form may be a problem for certain individuals who have limited writing skills, or who have a strong preference for verbal/face-to-face communication (see discussion of impact of culture in use of technology in Chapter 17). Second, people need to learn the importance of creating good subject headers and summarizing preceding discussions (or copying relevant portions of previous messages) to establish a proper context for understanding. Third, individuals need to be careful about how they express emotions via e-mail since this kind of expression can easily be misunderstood and trigger unwarranted arguments (i.e., "flaming"). Finally, e-mail must be read regularly by all parties for communication to work; a recalcitrant recipient will soon be ostracized and left out of the loop.

Actually one of the biggest potential problems with e-mail for many individuals and organizations is that it becomes too successful — everyone has too many e-mail messages to read and suffers from "information overload."

Groupware

Perhaps the most intriguing form of technology-mediated interaction is groupware — networked systems in which small groups work on-line. Groupware programs are designed to make it easy for people to create ideas or make decisions together even though they are physically separate (Baecker; Diaper and Sanger). In a typical groupware scenario, a group of people will sit at their computer workstations and enter their responses to various questions or issues. All of the responses entered can be seen by everyone involved on their screens with no indication of who provided what response. This allows people to respond anonymously without having to be concerned about contradicting others. In some systems, numerical responses can be automatically presented in terms of averages, medians, or some form of weighted score. Regardless of whether the results are qualitative or quantitative, a groupware environment allows all involved to express their views and consider those of others without precipitating arguments or ill-will. Such systems are very effective in financial decision-making (e.g., budget) and policy-making situations where each participant has vested interests.

On the other hand, some groupware systems are designed to make the participant's responses quite explicit. For example, each participant will have an ID tag associated with his/her responses; or each person may have his/her own window on the screen with his/her picture, icon, or actual video link. The key element of a shared workspace is still the focus of the system, but in this type of groupware, the identity of participants and their individual contributions is preserved. Such systems are popular for design and engineering work and usually feature graphics-creation capabilities at each computer workstation.

As with the other forms of technology-mediated interaction, learning to work together via groupware takes considerable experience before it becomes

---

The Liberator:
Assistive Technology in the Workplace

For individuals with cognitive disabilities such as autism, cerebral palsy, or multiple sclerosis, communicating with others is often a major obstacle. However, there are many forms of assistive technology that can help individuals overcome this obstacle sufficiently well to perform any job, no matter how demanding. One such device is the Liberator, a speech-generation system that allows those with speech impairments to talk and be understood without any problems. The product is distributed by Prentke Romich Co.

Bob Williams, one of the first severely disabled individuals to hold a major federal office (assuming you don't count President Franklin Roosevelt) has demonstrated how a device like the Liberator can do this. As Commissioner of Administration on Developmental Disabilities in the U.S. Department of Health and Human Services, he was responsible for managing a $110 million outreach program. Williams, who has had cerebral palsy since childhood, uses the Liberator to overcome his difficulties speaking — an activity which represents a significant part of his job. Of course, use of the Liberator is only part of the solution to William's adaptive work environment, which also includes a custom-made desk, specialized software tools, and a service dog.

Williams has demonstrated through his courage and persistence that disabled individuals can handle very demanding positions provided their employer is willing to provide/support the necessary adaptations and assistive technology.

From "People in Motion," a PBS television series that profiles how technology can help disabled individuals. Produced by WNET(New York) and available on videocassette.

---

successful — people may not trust the results of groupware sessions at first or they may feel that the quality of the responses is not as good as in traditional face-to-face meetings. But once participants get used to interacting this way, groupware sessions can be highly productive and free from some of the problems associated with personal interaction.

Accommodating Special Needs

One aspect of technology use in organizations that deserves special mention is the use of technology by individuals with disabilities. Various laws, most notably the Americans with Disabilities Act (ADA) passed in 1990, require that employers accommodate the needs of handicapped individuals in the workplace and elsewhere. These regulations apply as much to technology as anything else. Indeed, from a business perspective, it is very much in the interests of a company to ensure that all its employees are as productive as possible, and that all customers can be served fully.

Two other aspects of technology that need to be addressed are 1) providing individuals with adaptive or assistive devices needed to do their jobs or avail themselves of services, and 2) ensuring that any technology used by an organization can be used properly by people with disabilities (Covington and Hannah; Lazzaro). Actually, if technology is designed correctly, it should be usable by anyone and hence substantially reduce the need for specialized assistive devices. For example, it has become increasingly common to provide multiple modalities in software interfaces, e.g., a word processing program that can read the text that is displayed in a file. This eliminates the need for a blind individual to have to provide his or her own text-to-speech program. Similarly, developers of operating software now provide a set of basic functions such as screen magnification or "sticky keys" (allowing sequential instead of simultaneous key presses) to accommodate certain kinds of disabilities.

On the other hand, there are some types of assistive technologies that need to be customized for specific individuals. A typical category is input devices that allow people with mobility limitations or no arms to use a keyboard or mouse. It is also common for people with disabilities to need furniture that allows them to access technology from a wheelchair or in different positions. Problems also arise when multimedia programs used are not accommodated to individuals with disabilities. For example, a program that relies heavily on visual images, animation, or video will pose a dilemma for a blind person unless the program provides a text or spoken description of the images/video sequences. Similarly, a program that relies heavily on narration or sound is going to be a problem for a hearing-impaired individual unless there is an accompanying text description of the audio. So, CD-ROMs or web sites that involve extensive multimedia presentations need to offer information in multiple ways or adaptive devices will be required to overcome the limitations imposed on the disabled.

There is a simple way to ensure that these kinds of problems don't arise: involve disabled individuals and their representatives in all technology design and selection efforts as a matter of routine.

Who Can You Trust? Piracy, Privacy, Security, and Risk

While not directly related to training, there are a number of ethical issues that affect all technology-based activities in an organization and they should be mentioned. Indeed, teaching people about the ethical aspects of technology is an important category of knowledge that is neglected or overlooked in too many organizations. This includes piracy, privacy, security, and the risks associated with technology use.

Piracy refers to illegal copying of software, or the unauthorized use of material from computer or on-line sources (i.e., copyright violation). Both of these activities cost software companies and electronic publishers billions of dollars per year. In many organizations and educational institutions, employees and students make illegal copies of programs they find on computers for their own personal use. While this is a possibility with all media, it is especially easy to do with electronic media since it only takes a few seconds to copy something from one machine to another on a network (or to a floppy from a hard disk). Most organizations and institutions try to control this problem by obtaining licenses that cover use by all employees or students, but this usually doesn't include making copies for home machines or friends. Another way to control the problem is to "lock" hard disks (especially those on networks) so that programs cannot be copied. Copyright violations are harder to control because it is harder to prevent people copying information from files. Furthermore, in the case of on-line databases and the worldwide web, the information is outside the domain of the organization and belongs to the provider. While there are various ways to limit access to information, most people want to encourage as much access as possible, and hence restrictions are not popular.

The real solution to both of these piracy problems is not limited access (although there are certainly legitimate cases where this is desirable), but education of all computer users on the ethical considerations of electronic information — that copying programs and unauthorized use of information is stealing just as much as with physical items. Organizations such as the Software Publishing Association (SPA) and the Association for Computing Machines (ACM) are very active in promoting this message, but it is necessary for every business and institution to do so as well. This is one of primary functions of senior management in the information age: to establish and promote good ethical behavior on the part of its employees or members with respect to technology use.

There is another side of the coin — ensuring privacy of electronic information, especially when it pertains to employee, member, or student databases. Every organization keeps considerable information about its employees or members, much of which must be turned over to government or private agencies for processing taxes or benefits. Considerable lengths are usually taken to ensure that this information remains confidential, but sometimes there are "leaks" or errors that occur that compromise the privacy of information databases. Individuals have a right to be reassured that all reasonable precautions have been taken by organizations to ensure that such problems are very unlikely to happen. In fact, in some countries, there are specific laws pertaining to this right to privacy of information databases, although at the present time there are none in the U.S.

The issue of privacy is particularly relevant to the use of technology for learning because it is relatively easy in most systems to track what the user is doing and keep records of on-line activities. In fact, this capability for on-line tracking is one of the selling features for computer-based instruction — student progress can be monitored automatically and unobtrusively, with detailed reports available for the instructor or training manager. Many organizations have used this capability to determine the performance levels of employees who work at computers and determine if specific training activities (or other interventions) are effective. Unfortunately, some organizations have also monitored the e-mail messages and on-line interactions of their employees to determine if they are work-related. While organizations are within the (current) law to do this, it seems an affront to the privacy of their employees. At the very least, all individuals should be warned that their on-line actions are being recorded and may be examined.

Security of on-line information is another topic related to privacy — how well individuals and organizations are able to keep unauthorized people from accessing (and in some cases, abusing) computer systems. Much publicity has been afforded to so-called "hackers" (or more accurately, "crackers"), who illegally access networks and systems for the purpose of fraud or just mischief. The problems caused by computer viruses are also well known and very disruptive. Every computer user needs to know the basic steps to take to minimize security violations such as frequently changing their account passwords (and not revealing them to others), or running a virus detection program on all machines. Indeed, passwords are a fundamental form of security for many forms of technology (e.g., ATM systems, voice mail accounts, building access), and the same precautions apply.

This discussion about privacy and security considerations brings us to the last, and perhaps most profound, ethical issue concerning technology — how much can it be trusted? A number of very knowledgeable scientists (e.g., Kling; Landauer; Newmann; Norman) have documented how often things fail due to poor design or implementation. The public, and most employees, seem to have implicit trust that technology is highly dependable and accountable. But the truth is otherwise. Most large-scale computer systems and technology efforts are usually quite fragile and prone to failure. Systems that have to be reliable, such as transportation, financial or medical, are redundantly engineered so breakdowns have little or no impact on their operations. But most systems developers feel they cannot afford redundant design (or the extensive testing that would identify potential problems), so we live with technology that is highly likely to fail and produce mistakes.

This means that when we rely on technology for learning and training, we have to accept the fact that some of the time it's not going to work properly and occasionally will go haywire. Most instructors who work extensively with technology learn to be flexible and always have contingency plans. So if a session was to be conducted via satellite teleconference, but the link is down, it can still be done via telephone. Or if materials were to be distributed via the web, and the server has problems, there is still fax or maybe overnight couriers. Still, this can be disruptive to learning and often prevents educational activities from proceeding as intended. Who is to blame? What kind of impression/attitude do students, employees, or customers develop about technology-based learning? Is tolerating technology failure the same as dealing with people who are absent due to illness? What are the ethics of systems that don't work?

Technoculture and Technophobia

Finally, we should mention that there are individuals who can't or won't accept technology as a generally beneficial thing. In every organization and institution, there will be some employees and students who don't want to learn via technology, nor use it in their jobs or lives. Or they may be willing, but simply can't get the hang of it. According to recent surveys, up to 70% of Fortune 1000 companies say the success in using technology has been limited by the low interest and skill level of their employees.

Although it would be pleasant to say that individuals who don't want to use technology should not be forced to do so, the fact remains that in almost any workplace scenario it may not be possible to offer this option. For example, if an organization relies on a computer system to process orders or sales, it is unlikely that employees or customers can do their transactions manually (although a manual procedure ought to exist for back-up purposes). However, it may be possible to offer choices in the way computer tasks are conducted or perhaps multiple hardware/software alternatives (which may also address the needs of disabled users).

Why don't employees use the available technology for learning and accessing knowledge? There are many reasons. In some cases, the resistance to using technology is more of an objection (sometimes quite justified) to a lack of flexibility in how job tasks are carried out and the rigidity of computer procedures. Allowing individuals to customize the hardware, software, or system methods to suit their needs or preferences may significantly reduce opposition to its use.

In other cases, objections to technology may be more fundamental — concerns about how the use of technology has made a task or job unfulfilling or too tedious. Or the use of technology may make a job too complex and overwhelming, relative to its past form. Technology can, in fact, make tasks more abstract and conceptual compared to their manual counterpart. For example, producing documents using word processing and desktop publishing is more cognitive in nature when compared to the past use of typewriters and typesetting equipment. Similarly, working in an automated factory with robotic or computer-controlled machines is much different than when all assembly was done by hand. And, as we discussed earlier in this chapter, interaction is increasingly mediated by technology rather than in-person, which is difficult for some individuals to accept and be comfortable with.

There are also a number of other reasons why technology is avoided or receives limited use in the workplace.

1. Individuals are not adequately trained and motivated to use available technologies, sometimes lacking even basic computer skills to access the system.
2. Search time associated with finding needed knowledge is seen as inconvenient and excessive relative to the value gained.
3. Many are not able to analyze their problems in a way that allows technology to contribute the viable solutions.
4. Selecting knowledge from a technological database may alert others to an individual's weakness and needs.

5. Contributing knowledge to a knowledge database may be seen as weakening one's personal power.
6. There is little or no reward or recognition for contributions to the organization's database, and therefore energies are not devoted to that effort.
7. The technology-based learning system may compete with the many formal and informal systems for exchanging knowledge across units which exist.
8. There is an absence of a learning culture so that knowledge creation and sharing is not encouraged.

It is important to address all these reasons for rejecting technology because ultimately people are needed to make the technology work and be of value. Like many of the technologies developed over the years (e.g., electricity, automobiles, telephones), computer-based systems are part of this historical and evolutionary trend. We must help people see how technology enables them to be more productive and satisfied with their lives. The ability to work with technology to find information and convert it into knowledge has become an essential skill of every worker. Ultimately, technology changes the way we learn and teach in the workplace. In Part II, we will examine how technology assists us in learning and training.

References

Baecker, R.M. (1993) *Groupware and CSCW.* San Mateo, CA: Morgan Kaufman Publishers.

Covington, G. and Hannah, B. (1996) *Access by Design.* New York: Van Nostrand Reinhold.

Diaper, D. and Sanger, C. (1993) *CSCW in Practice: An Introduction and Case Studies.* New York: Springer-Verlag.

Dreyfus, H. (1992) *What Computers (Still) Can't Do.* Cambridge, MA: MIT Press.

Galegher, J., Kraut, R., and Egido, C. (1990) *Intellectual Teamwork: Social and Technical Bases of Collaborative Work.* Hillsdale, NJ: Erlbaum.

Kling, R. (1995) *Computerization and Controversy: Value Conflicts and Social Choices* (2nd Ed). San Diego, CA: Academic Press.

Landauer, T. (1995) *The Trouble with Computers.* Cambridge, MA: MIT Press.

Lazzaro, J. (1996) *Adapting PCs for Disabilities.* Reading, MA: Addison-Wesley.

Neumann, P. (1995) *Computer Related Risks.* New York: ACM Press/Addison-Wesley.

Norman, D. (1993) *Things That Make Us Smart.* Reading, MA: Addison-Wesley.

Postman, N. (1992) *Technopoly.* New York: Knopf.

Rosenberg, R. (1997) *The Social Impact of Computers* (2nd Ed.) San Diego, CA: Academic Press.

5 Learning Technologies for Maximizing Human Performance

At the Boeing Training Center in Seattle, Washington, pilots and ground crew technicians are trained via integrated CD-ROM and laser discs. Training modules are downloaded into the network. Each training carrel includes a Windows-based PC, where laser animation and audio can be overlaid on a computer-generated graphic to stimulate changing scenery or weather conditions outside a cockpit "window." Other programs simulate pilot–copilot interaction for flight training, modules so realistic that pilots can be certified for flight without ever logging a minute in the air. According to Jack Hide, Boeing computer training technologist, this form of technological learning achieves "maximum training flexibility in a constantly changing environment" (Benson and Cheney). The old way of using trainers and training manuals is too cumbersome, unreliable, and expensive in today's competitive marketplace. Technology "simply does the job faster and better."

Welcome to workplace learning for the new millennium! Where training will be just-in-time, just-what's-needed, and just-where-it's needed; customized, digitized, and optimized for each individual — thanks to the ever-increasing power of technology as well as its growing applicability to learning in the workplace.

Surge in the Use of Learning Technologies

One of the most striking phenomenon occurring in the workplace as we enter the 21st century is the unrelenting demand for increased knowledge and speed of learning. A small but growing number of companies worldwide are recognizing that technology is the key to achieving this goal of better and faster learning and knowledge. Respondents to a 1997 HRD Executive Survey reported that although only 10% of their organization's training time in 1996 was delivered by learning technologies, they expected that figure to soar to over 35% by the year 2000!

The types of technologies that are being used are quickly diversifying. Companies are now investing in a wide variety of the electronic learning technologies. Computer delivery of training, use of Internet and network-based distance learning, EPSS, and use of interactive/multimedia computer-based training surging. Organizations now spend over $4 billion a year on software for the intranet. According to the 1997 ASTD Survey, dramatic increases in all forms of learning technologies are occurring (Table 5.1). Technologies growing fastest are the intranet, Internet, and computer-based training.

Table 5.1   Use of Electronic Learning Technologies

Learning Technology	Percentage Using Technology in 1996	Percentage Expecting to Use Technology in 1997	Rank in the Year 2000
CBT: disc/hard drive	55.2	63.5	9
Video-teleconferencing	53.1	56.3	5
CBT: CD-ROM/CD-i	42.7	54.2	10
Interactive television/video	37.5	42.7	6
Multimedia: CD-ROM/CD-i	29.2	37.5	7
Internet/Web	27.1	47.9	3
CBT:LAN/WAN	21.9	41.7	4
Computer teleconferencing	14.6	22.9	8
Intranet	13.5	44.8	1
Multimedia: LAN/WAN	12.5	24.0	2
EPSS	4.2	13.5	11
Virtual reality/electronic simulation	1.0	2.1	12

Source: ASTD's National HRD Executive Survey, 1997

Types of Learning Technologies

Learning technology is "the use of electronic technologies to deliver information and facilitate the development of skills and knowledge." It includes both *presentation* (how information is presented to learners) and *distribution* (how information is delivered to learners) elements. Let's look at the various presentation and distribution learning technologies as categorized and defined by the American Society for Training and Development (ASTD).

Presentation Technologies

Electronic text or publishing — the dissemination of text via electronic means

CBT — learning that uses computers to deliver training

Multimedia — computer application that uses text, audio, animation, and/or video

Television — one-way video combined with two-way audio or other electronic response systems

Teleconferencing — the instantaneous exchange of audio, video, or text between two or more individuals or groups at two or more locations

Virtual reality — a computer application that provides an interactive, immersive, and three-dimensional learning experience through fully functional, realistic models

Electronic performance support system (EPSS) — an integrated computer application using expert systems, hypertext, embedded animation, and/or hypermedia to help and guide users to perform tasks

Distribution Technologies

Cable TV — the transmission of television signals via cable technology

CD-ROM — a format and system for recording, storing, and retrieving electronic information on a compact disk that is read using an optical drive

Electronic mail — the exchange of messages through computers

Extranet — a collaborative network that uses Internet technology to link organizations with their suppliers, customers, or other organizations that share common goals or information

Internet — a loose confederation of computer networks around the world that are connected through several primary networks

Intranet — Internets within an organization

Local area network (LAN) — a network of computers sharing the resources of a single processor or server within a relatively small geographic area

Wide area network (WAN) — a network of computers sharing the resources of one or more processors or servers over a relatively large geographic area

Satellite TV (also called business TV) — transmission of television signals via satellites

Simulator — a device or system that replicates or imitates a real device or system

World Wide Web — all the resources and users on the Internet using Hypertext Transport Protocol (HTTP)

In later chapters in this book, we will be describing and illustrating how the technologies listed above assist workplace learning (Part II) and how they assist knowledge management (Part III). Let us first, however, examine the benefits of learning via technology and how learning technology has affected the field of human resource development (HRD).

Advantages of Using Learning Technologies

Technology is becoming more popular because it increases the quality, relevance, and speed of learning. Among the advantages of learning technology:

1. Available as Needed and Just-in-Time

Unlike traditional training, which is just-in-case, this type of training enables individuals to find and use the information they need when they need it, available just-in-time. As Lew Parks, Vice President for Learning and Professional Development of AMS, notes, "People don't have to wait for class to be offered. Learning is ready when they are ready." Flexibility is such that the learning content can be an "ongoing work in progress." Employees have greater freedom to initiate the types of learning experiences they need to achieve improvements in their jobs.

2. Learner-Controlled

The focus of this new technology-based learning is under the control of the employee, often available at his/her worksite rather than in a distant corporate classroom. More and more courses are being automated using on-line documentation systems. Employees can learn from technology or self-guided learning workbooks rather than from central human resource development offices. Why is it so important for employees to have so much more control over their learning programs? The reason is that most jobs within the corporation are becoming ever more complex and require higher levels of skills, about which only the worker is fully cognizant. In addition, they know when they need it and for what purposes.

3. Cost-Effective

An important advantage of technology-based learning is the dramatic cost savings. Bradon Hall, editor of *Multimedia Training Newsletter,* predicts that technology will reduce the time required to deliver training by "30% to 60%." Savings occur because of the ability to use fewer instructors to reach many more participants, reduced travel expenses, reduced down time (when people might otherwise have to leave their offices — see FORDSTAR example in Chapter 18), and the ability to train a dispersed work force at the same time (synchronous) or different times (asynchronous). There is also growing ease and simplicity in developing and maintaining the learning packages, databases, and intranet sites with a minimum of cost and time.

4. Self-Paced and User-Friendly

Being able to easily access information and acquire skill is essential so workers are not discouraged from using the technology (see section on technophobia in Chapter 4). The learning technologies, as will be illustrated in the remaining chapters, are becoming easier to access and simpler to manipulate. Intranet applications, for example, typically use an interface that is conducive to simple point-and-click navigation.

5. Accessibility Over a Wide Geographic Area via Distance Learning

One of the greatest benefits of learning technology is its ability to provide distance learning where the instructor and/or training source is some distance (a few hundred feet or 10,000 miles) from the learners. Learning via technology allows accessibility to any number of geographical areas. Self or group training can be undertaken at home, in the office, or on the road. In determining which distance-learning technology to employ, a number of elements should be considered:

a. Be sure to gather a cross section of learners' demographic, education, and socioeconomic profiles before choosing a distance-learning technology.

b. Determine students' motivation — a highly motivated student population is likely to need less person-to-person monitoring than a population with little or no motivation for learning

6. Hands-On Interactiveness

Technology such as touchpads, Internet, and intranets allow for hands-on, direct, and immediate interaction with instructors and/or fellow learners. Groupware, chatrooms, and other two-way communication tools can also be easily integrated into the learning programs.

7. Uniformity of Content and Delivery

There is improved consistency since the same program is being offered by the same person and/or system throughout the company. The centralization of information and databases also assist to assure this capability.

8. Adjustment to Individual Learning Styles

It is important to take into account the way learners naturally learn and think (multiple learning styles; multiple orientations such as visual, aural, kinesthetic; multiple preferences for problem solving). There are a number of models that have been developed to classify learning styles, but among the best is Honey and Mumford's classification:

Activist — This learner involves self fully in new experiences, enjoys the here and now, is open-minded, and is bored with implementation and longer-term consolidation.

Reflector — This person stands back and ponders experiences from many different perspectives, collects data and thinks before reaching conclusions, values analysis, is cautious, thoughtful, and listens before commenting.

Theorist — This individual is strong at adapting and integrating observations, logical, uses a step-by-step approach, is a synthesizer, systems thinker, and uses rational objectivity.

Practitioner — This learner is an experimenter who is interested in trying out new ideas, theories, and techniques, acts quickly on ideas that are attractive, practical, and sees problems as a challenge.

The wide variety of technologies allows us to develop programs that appeal to each and all of these learning styles, sometimes even at the same time. For example, artificial intelligence, which involves replicating the thought process of the human brain, can observe, guide, and coach users and modify its instructions accordingly. It can adapt to each user's cognitive style, resulting in customized help that corresponds to the needs of each trainee. Training is thus faster, more interesting, more applicable, and more motivating because it introduces only information needed by the user.

9. Adjustment to Motivation Level of Learners

Technologies can also be adapted to the motivation levels of the learners. The best technology for motivated learners appears to be video, self-paced workbooks, audiotapes, and CDs. For these more enthusiastic participants, the program design can be self-paced, the materials need not be glitzy, the content need only be self-explanatory, and the package should contain self-help materials. The less motivated learners prefer learning technologies such as videoconferencing, cable television, two-way audio, satellite broadcasts, interactive CD, and kiosk systems. In designing programs for these students, there needs to be monitoring and more interaction with each other and the instructors. Attendance may need to be mandatory, and the content can be expanded and varied based on audience questions.

10. Safety and Flexibility

Using virtual reality technology, for example, to deliver training modules is especially useful when it allows trainees to view objects from a perspective that would be impractical or impossible in reality. For example, it is not practical or safe to turn a drill press on its side so you can see the bottom as a front view. Cyber-training also has applications beyond manufacturing or traditional blue-collar tasks; for example, brokerage firms now use virtual reality on the job in real time to train brokers.

11. Ability to Continuously Update

Technologies, especially those available via the Web, allow organizations to easily update and make available information to all employees. Intranet sites can be also easily and inexpensively amended as frequently as needed.

12. Availability of Both Push and Pull Approaches

Too often, employees are provided with more information than they can possibly process or retain. Technologies like the intranets allow companies to provide access to as little or as much information as employees wish to pull onto their desktops.

Technology Accelerates Transformation from Training to Learning

For thousands of years the traditional way of education and learning was to bring the learners together in a single location with lecturers, lectures, and books. This is now rapidly changing. The traditional way of teaching and training has become much too slow, ineffective, and expensive in today's highly competitive workplace.

Why? What's wrong with training? Almost everything! First, let's look at the amount of knowledge that actually gets transferred in a typical training program. Research shows that less than 15% of the material covered in the classroom or corporate training room ever gets applied to the job (Broad and Newstrom). There are a number of reasons for this:

1. The training is provided to a group of people with varying degrees of interest and expertise in the subject matter at the time they are receiving the training.

2. Most of the content is being provided on a just-in-case basis; many participants are not sure if or when they will need to apply the information, and therefore make less effort to learn.
3. Some of the material may be too advanced or too easy for the participants, and therefore be above their abilities (frustration) or below their abilities (boring).
4. Since instructors are reaching a relatively small number of people (5 to 30), they do not/cannot take the time nor do they have the financial or technical resources to design and deliver a top-quality program.

For these reasons a lot of the learning never reaches the participants before they can even reject or forget it. Learning, on the other hand, differs in several significant ways from training:

1. In *training* someone else is responsible for seeing that you acquire the competencies you need. In learning *you* are the one responsible. In today's rapidly changing environment and organizational structures with self-managed teams or managers supervising 30 to 300 people rather than 4 or 5, the individual is the only one who can stay abreast of his/her learning needs and identify resources to provide such learning.
2. Much of training is just-in-case, at this time, with this group, by this instructor. Learning is just-in-time, anytime, and anywhere. You learn something that you will need to apply in the near future, and we all pay much closer attention and give more energy to something we will need to perform soon and for which we will be held accountable.

Although there will always be a need for training, the amount of time organizations spend in providing classroom training programs will drop from the current 80% to 20% within the next few years.

Managing the Learning Technologies

As we move from training to learning, a key role of the HRD administrator will be to manage the transition from classroom training to distance learning, be it via interactive multimedia, teleconferencing, CBT, or web delivery systems such as Internet and intranet. Support electronic technologies will be available for course design, production, and management, including software applications that assist with word processing, multimedia authoring, database management, performance management, logistic management, and course registration.

Technology will raise a number of unique challenges for HRD managers and administrators such as:

1. How to train the large numbers of temporary, contact, or at-home workers
2. How to deal with the need to have more reliance on outside partners, and the sharing of training production
3. Determining how to manage the costly aspects of some of the learning technologies, since they may be expensive to purchase, develop, and maintain
4. How to update data, software, and hardware

Changing HRD Roles Due to Profusion of Technology

The roles and activities of HRD staff have become transformed in a variety of ways as more and more technology enters the workplace. The traditional, centralized corporate training department, with its catalogs of classes and workshops, is becoming a relic (Shandler; Brinkerhoff and Gill).

Roles and Skills of HRD Staff

The corporate HRD professional is quickly evolving into someone who facilitates, mentors, and guides employers and employees to use the best and most timely training available. The goal of the corporate trainer has become to find, interpret, and assess a wide range of information and technologically sophisticated products. "Intersector directors" is the term Mantyla and Gividen use to describe these new roles. HRD professionals are now redirecting their thinking on how training can empower the company as a competitive leader. Trainers, instead of focusing on teaching and presentation skills, will rely more on instructional development skills and advancing learning technologies.

The Way Training Is Delivered

Classroom training will be largely replaced by individual learner-oriented training, delivered through interactive videos available at production/service-line workstations. Innovations in microchip and satellite technology, expanded computer memory, development of expert systems and artificial intelligence, the promise of simulations training for high-risk occupations, and advances in the production of interactive technologies will change significantly the availability and quality of teaching and learning, thus providing new, cost-effective delivery methods. Training delivery will continue its movement from professional trainers to nontrainers such as managers, team leaders, and technical workers.

New HRD Competencies Needed to Manage Learning Technologies

The American Society for Training and Development (ASTD) recently completed a survey of HRD technologists to list the technology-based HRD competencies needed in today's workplace. A total of 36 technology-specific competencies were identified under the categories of (a) general, (b) distribution method, and (c) presentation method.

A. General Competencies
1. *Awareness of technology industry:* having a general understanding of the trends within the learning technology industry and knowing the existing and emerging technologies
2. *Program evaluation:* measuring the success of technologies used in the delivery process
3. *Management of learning technology selection:* supervising the selection of learning technology or combination of technologies to meet specific needs; determining when, how, and where learning technologies should be used and monitoring the progress of all the other roles in the delivery process
4. *Management of learning technology design and development:* supervising and assuring the effective integration of performance objectives, course materials, and learning technologies into a design document that fulfills the organization's goals
5. *Management of learning technology implementation, support, and evaluation:* supervising the installation and maintenance of learning technologies and assuring that all systems continuously meet company specifications

B. Distribution Method Competencies

6. *Cost analysis/ROI of the distribution methods:* understanding the relative costs of each distribution method, and assuring that the organization is receiving a good value for the money spent on these technologies

7. *Limitations and benefits of the distribution method:* knowing the true capabilities of each technology and tying these in with the needs of the organization

8. *Effect of distribution method on learners:* assessing how various distribution methods will cater to individual learning styles; balancing learner needs against organizational needs

9. *Integrating distribution methods:* Mixing distribution methods in an effective and efficient manner to facilitate learning

10. *Remote site coordination:* coordinating the installation and maintenance of distribution technologies at a remote site and assuring that all systems continuously meet design specifications

C. Presentation Method Competencies

11. *Electronic text design and development:* outlining and creating text-formatted, instructional materials that are suitable for electronic dissemination; determining which instructional methods are best suited to electronic text and which distribution methods will best deliver the final program to the learner

12. *Electronic text implementation, support, and evaluation:* coordinating the installation and maintenance of distribution technologies that disseminate text-formatted materials and assuring that all systems continuously meet design specifications

13. *Computer Based Technology (CBT) design and development:* outlining and creating CBT-formatted (text based) instructional materials that are suitable for electronic dissemination; determining which instructional methods are best suited to CBT and which distribution method(s) will best deliver the final program to the learner

14. *CBT implementation, support, and evaluation:* coordinating the installation and maintenance of distribution technologies that disseminate CBT-formatted materials and assuring that all systems continuously meet design specifications.

15. *Multimedia design and development:* outlining and creating multimedia-formatted (text, graphics, and audio), instructional materials that are suitable for electronic dissemination; determining which instruc-

tional methods are best suited for multimedia and which distribution method(s) will best deliver the final program to the learner

16. *Multimedia implementation, support, and evaluation:* coordinating the installation and maintenance of distribution technologies that disseminate multimedia-formatted (text, graphics, and audio) materials and assuring that all systems continuously meet design specifications

17. *Interactive TV design and development:* outlining and creating interactive TV events that are best-suited to one-way video, two-way audio dissemination of information; determining which instructional methods are best suited for interactive TV and which distribution method(s) will best deliver the final program to the learner

18. *Interactive TV implementation, support, and evaluation:* coordinating the installation and maintenance of distribution technologies that allow interactive TV events to occur and assuring that all systems continuously meet design specifications

19. *Teleconferencing design and development:* outlining and creating teleconferencing events that are best suited to two-way video and two-way audio dissemination of information; determining which instructional method(s) will best deliver the instantaneous exchange of information between participants

20. *Teleconferencing implementation, support, and evaluation:* coordinating the installation and maintenance of distribution technologies that allow teleconferencing events to occur and assuring that all systems continuously meet design specifications

21. *On-line help design and development:* outlining and creating tools that provide immediate assistance to individuals and groups and are suitable for electronic dissemination; determining which instructional methods are best suited to on-line help, and which distribution method(s) will best deliver the final program to the learner

22. *On-line help implementation, support, and evaluation:* coordinating the installation and maintenance of distribution technologies that allow on-line help systems to operate and assuring that all systems continuously meet design specifications

23. *Groupware design and development:* outlining and creating tools that allow group collaboration for learning; determining which instructional methods are best suited to Groupware and which distribution method(s) will best deliver the final program to the learner

24. *Groupware implementation, support, and evaluation*: coordinating the installation and maintenance of distribution technologies that allow Groupware systems to operate and assuring that all systems continuously meet design specifications

25. *Virtual reality/3D modeling design and development*: outlining and creating instructional opportunities that include interactive, immersive, and three-dimensional learning experiences; determining which instructional methods are best suited to virtual reality and which distribution method(s) will best deliver the final program to the learner

26. *Virtual reality/3D modeling implementation, support, and evaluation*: coordinating the installation and maintenance of distribution technologies that allow virtual reality systems to operate and assuring that all systems continuously meet design specifications

27. *Audio design and development*: outlining and creating instructional materials that are effectively implemented through one-way delivery of sound; determining which instructional methods are best suited to audio and which distribution method(s) will best deliver the final program to the learner

28. *Audio implementation, support, and evaluation*: coordinating the installation and maintenance of distribution technologies that disseminate audio-formatted materials and assuring that all systems continuously meet design specifications

29. *Video design and development*: outlining and creating instructional materials that are effectively implemented through one-way delivery of live or recorded full-motion pictures; determining which instructional methods are best suited to video and which distribution method(s) will best deliver the final program to the learner

30. *Video implementation, support, and evaluation*: coordinating the installation and maintenance of distribution technologies that disseminate video-formatted materials and assuring that all systems continuously meet design specifications

31. *Electronic Performance Support Systems (EPSS) design and development*: outlining and creating an application system that effectively combines a variety of expert systems; deciding which systems (hypertext, embedded animation, hypermedia) will provide users with support necessary to perform a required task; determining which instructional method(s) are best suited to EPSS and which distribution method(s) will best deliver the final program to the learner

32. *EPSS implementation, support, and evaluation*: coordinating the installation and maintenance of distribution technologies that allow

EPSS to operate and assuring that all systems continuously meet design specifications

33. *Performance-centered design*: designing software to support high-performance work, using the best practices for the design of a human/computer interaction that provides job performers with learning, referencing, and other resources they need to do the work

34. *Desktop computer literacy*: using a desktop computer for designing, implementing, and evaluating learning technologies

35. *Programming*: using authoring tools to develop electronic instructional materials

36. *Software quality assurance*: determining if software is performing according to its specifications

Analysis and Assessment Skills

A common characteristic of many of the 36 competency areas described above is their requirement for analysis and assessment skills. While technology will likely automate many aspects of training system design and implementation, HRD professionals will still need to be able to conduct needs assessments, problem analysis, cost/benefit evaluations, etc. Indeed, given the complexity of knowledge management systems and EPSS, good "front-end" analysis studies will be paramount.

Needs assessment and job/task analysis are two of the most fundamental components of Instructional Systems Design (ISD) — a methodology that has been used in the training world for years — and the basis for the practice of Performance Technology. Needs assessment is concerned with determining why a performance problem exists; job/task analysis is focused on specifying, in as much detail as possible, what employees need to know (or be able to do) in order to perform their jobs. Even though the shift from the training to the learning paradigm places more responsibility on employees to determine their own needs, it is still necessary for someone to ensure that the necessary information and learning resources are available.

In the case of computer-based systems (and all future technologies are computer-based in one way or another), a great deal of user interface design is required (discussed further in Chapter 9). In the development of a knowledge management system, a great deal of time and effort is devoted to determining exactly what functions and features a particular group of users wants in order to perform certain tasks. This is a special case of job/task analysis in the context of computer system development. However, the development of knowledge management systems or EPSS requires a much more comprehensive analysis, because it normally involves many organizational interactions rather than an isolated set of tasks that only affect one person. In addition, such systems are commonly implemented in a global context, so there are cross-cultural factors to consider (see Chapter 17).

The important implication is that, even though HRD specialists will need to have a high level of technological sophistication in the future to do their jobs, they will also need to have high-quality analytical skills. Technology-based learning systems will not be effective unless they are based on accurate assessments of needs and requirements.

Exploring the Learning Technologies

New technologies and innovative use of them have enabled organizations to quickly leapfrog over competitors. The beauty of marrying interactive technology with learning has resulted in greatly improved retention, increased flexibility, and greatly enhanced learning. In this chapter, we examined the types of learning technologies, how technology empowers workers in the workplace, and the changing HRD roles and competencies caused by technology. In the upcoming chapters of Part II, we will look in greater detail at the following learning technologies: electronic publishing (Chapter 6), television and video (Chapter 7), teleconferencing (Chapter 8), interactive multimedia (Chapter 9), simulation, simulators, and virtual reality (Chapter 10), and authoring (Chapter 11).

References

American Society for Training and Development. (1997) *Learning Technology Survey.* Unpublished.

Bassi, L., Cheney, S., and Van Buren, M. (November, 1997) Training Industry Trends 1997. *Training and Development.*

Benson, G. and Cheney, S. (October, 1996) Best Practices in Training Delivery. *Technical and Skills Training.*

Brinkerhoff, R. and Gill, S. (1994) *The Learning Alliance: Systems Thinking in Human Resource Development.* San Francisco: Jossey-Bass, 1994.

Broad, M. and Newstrom, J. (1992) *Transfer of Learning.* Reading: Addison-Wesley.

Honey, P. and Mumford, A. *Learning Styles Inventory.* King of Prussia, PA: HRD Press.

Mantyla, K. and Gividen, J. (1997) *Distance Learning.* Alexandria, VA: ASTD Press.

Marquardt, M. (1996) *Building the Learning Organization.* New York: McGraw-Hill.

Shandler, D. (1996) *Reengineering the Training Function.* Delray Beach, FL: St. Lucie Press.

6 | Electronic Publishing

While we will discuss a range of different audiovisual, multimedia, and network technologies in subsequent chapters, the technology with the most impact on information flow in organizations and the development of learning programs is still "print." However, the way print documents are created and distributed has changed tremendously over the past few decades and promises to change even more in the coming years. In this chapter, we examine the fundamentals of electronic publishing and how it affects learning programs and activities.

Doing It on the Desktop

Prior to the widespread availability of personal computers and laser printers, the process of producing documents was primarily confined to printing departments (often outside companies) who had the necessary equipment and skills required to use them. Such groups also tended to have graphic arts specialists, who developed the artwork (i.e., illustrations, page layouts, photographs) for publication. Control of document production was localized in these departments, and creating publications involved a significant amount of scheduling as well as bidding and quality-control steps.

Desktop publishing has changed this equation dramatically. With the emergence of PCs, inexpensive laser printers, and page layout software (e.g., Adobe PageMaker, Corel Ventura), it became possible for everyone to produce professional-looking documents. No longer was it necessary to go through a long production cycle that involved the printing department — many types of documents could be done quickly and inexpensively using desktop publishing technology. Indeed, over time this technology has become even more flexible with the advent of page layout features in word processing software (such as Word Perfect or Word), color printers, scanners, digital cameras, and various kinds of graphics programs (including electronic clip art and digital photo libraries).

Exactly what were the implications of this amazing development for training and HRD activities? Training and HRD groups produce a tremendous number of publications: guides, manuals, memos, forms, newsletters, questionnaires, brochures, directories, outlines, and catalogs, as well as other documents. Under the old system of printing departments, the production of all these documents was a major management headache. With desktop publishing systems, the production of documents can be localized to the group needing them. In other words, desktop publishing technology resulted in a significant decentralization of power within an organization with regard to information flow. Each group was free to manage the production process for their own documents, including scheduling and design decisions.

This new-found power extends all the way to the individual. With desktop publishing systems on their desks, employees can produce professional-looking documents (company reports or their resumes) with no need to involve anyone else in their efforts. Small business owners and consultants can now produce documents of the same quality as big corporations. Even kids can create impressive papers to turn in for class assignments. So desktop publishing is a great equalizer among individuals relative to the ability to turn out high-quality materials in a short time.

While this decentralization is overall a good thing, there are some serious issues and problems to be addressed. Remember that under the old system, there were printing and graphic arts experts who decided how documents were composed. In the new desktop publishing regime, such decisions need to be made by every person who uses a page layout or graphics program. Suddenly thousands of people now need to understand basic layout and graphics principles (e.g., What is a font, anyway?). It was discovered that some people did or did not have the patience or predilection to compose documents. Books, seminars, and consultants on desktop publishing proliferated. Composition software started to include templates that provided a selection of completed designs. Many of the documents produced by desktop publishing were definitely inferior to the work previously done by printers and graphic design experts — at least in the beginning.

A second consideration is whether having each group publish its own documents is efficient from an organizational perspective. In terms of labor, equipment, and materials costs (think of all that wasted paper!), it probably is not. Furthermore, with the capability to produce documents so easily available, desktop publishing technology created an explosion in the number of documents produced. Now every group in an organization can produce its own manuals, guides, newsletters, forms, etc. Having a central publishing group allowed the organization (either explicitly or surreptitiously) to keep the number of publications limited and minimize redundancy.

A third issue is the need for so many people to learn how to use desktop publishing hardware and software. While laser printers and page composition programs are not difficult to learn, they still involve a significant training effort. Teaching people how to use desktop publishing systems has become a major training task in most organizations and another aspect of being computer literate. Furthermore, this training is not a one-time, thing because the systems and software continually change as new features and functions are added.

Not all publications are suitable for desktop publishing efforts. Documents that are very large (e.g., catalogs or directories) or very fancy (e.g., glossy magazines or brochures) are better done by traditional printing departments using high-speed presses. What desktop publishing technology has done is to reduce the workload of printing departments to larger, more complex jobs and leave the production of simpler documents to hundreds of other people in the organization. However, even this situation has been changed by electronic publishing technology.

The Electronic Document Pipeline

Up to this point we have mainly been describing the final step in the production of a document: page layout and printing. But for many kinds of publications, much of the effort takes place before this step, i.e., in the creation and collection of the information itself (Figure 6.1). For example, in a typical training manual for a product or piece of equipment, a great deal of time is involved in determining what information needs to be included (so-called "front-end analysis") and then having the appropriate people generate that information. This information then needs to reviewed and edited — and perhaps reorganized/rewritten by someone other than the original writer. All of this activity can take place in electronic form via shared documents on a network and/or e-mail messages. Indeed this is one of the primary applications for a Local Area Network (LAN) in any organization — to allow the creation of documents across a group of different individuals.

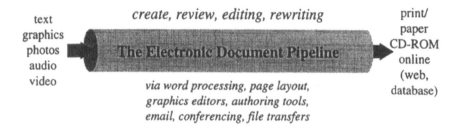

text
graphics
photos
audio
video

create, review, editing, rewriting

The Electronic Document Pipeline

*via word processing, page layout,
graphics editors, authoring tools,
email, conferencing, file transfers*

print/
paper
CD-ROM
online
(web,
database)

Figure 6.1   The Electronic Document Pipeline

So even documents that are to be published in a more or less traditional fashion can be composed electronically. What is delivered to the printing department (or printing company) is the final version of the document as a computer file. Alternatively, a publications group may take responsibility for assembling the individual documents in electronic format from various contributors and then use page composition software to create the final publication before turning it over to a printing department. This is essentially what most modern magazines and newspapers do. They receive articles as computer files from writers and then edit and assemble them electronically. Photos and graphics can likewise be submitted in electronic format and edited/assembled into the final publication. The interesting thing about this electronic pipeline is that contributions can be coming from someone down the hall via a LAN or from a person on the other side of the world using the Internet — but there is no distinction as far as computer files are concerned; they all look identical.

This electronic composition of publications has a further wrinkle in the distribution process. In traditional publishing efforts, documents are printed in one location and then physically shipped to their end destination. But it is often possible to electronically transmit the document in final form to its destination where the required number of copies are then printed out. This is called "on-demand" printing. It can involve a regional printing plant for a magazine or a high-capacity laser printer at a branch office. On-demand printing has a number of benefits that include elimination of the costs and time delay of shipping, as well as the capability to print just enough copies of a document for current orders or needs. The critical factor in making this kind of publishing work is having the right kind of printing equipment (hardware and software) at the destination locations.

Another form of on-demand printing is the simple distribution by fax machines. Many organizations (both corporate and government) have set up fax request systems that allow you to request a particular document via a phone call and have it immediately transmitted to your fax machine. In addition, it is possible to "broadcast" documents via fax — a technique that is used by many organizations to provide important news bulletins to their employees or customers. Most fax machines are now capable of sending the same transmission to many numbers simultaneously. As fax machines become increasingly prevalent in offices and homes, this is a viable way to distribute certain kinds of documents.

The Technology of Text

At the same time the hardware and software for electronic publishing has been evolving, there has been an increasing awareness and interest in the underlying principles and processes involved in creating print documents, especially the kind of technical publications used in training and HRD domains.

In creating a document (whether on paper or in electronic form), there are many design decisions to be made about type size/styles, spacing, use of headings, paragraph structure, page layout, and the positioning of graphics relative to text. Furthermore, there are many techniques of good writing, language usage, and information organization that make a significant difference to the readability and effectiveness of documents. These techniques, along with the research that supports them, are described in a number of books about the technology of text (e.g., Gropper; Jonassen; Misanchuk). While most of these principles are commonly known (e.g., use short sentences and the active voice; break material into small units with subheadings, minimize the number of different type styles used, etc.), they are not commonly practiced, resulting in documents that are difficult to read and understand.

In addition to these general guidelines for creating text publications, there are also some theoretical frameworks for the design and organization of documents. One very influential framework is the structured writing system developed by Robert Horn (also known as information mapping). The essence of the structured writing approach is the organization of content into "information blocks" (in contrast to traditional paragraphs) governed by four basic principles: chunking, labeling, relevance, and consistency. Structured writing has been applied to the design of many publications at different organizations with well-documented results (Horn, 1989; 1992).

Another important design framework for technical publications is the minimalism methodology developed by John Carroll. This methodology is specifically aimed at the design of computer manuals and focuses on what information should be presented and how it should be organized. What makes this theoretical framework particularly interesting is that its guidelines are somewhat contradictory to traditional instructional design principles. For example, minimalism de-emphasizes the importance of specifying detailed steps as a procedure in favor of allowing the person to explore his or her own path to achieve a well-defined goal. Minimalism also places a lot of emphasis on error handling and the relevance of the content to the system being learned.

More Than Words

Of course, there is a lot more to most documents than just text. Many publications (especially those intended for education purposes) have illustrations, diagrams, schematics, photographs, graphs, and/or charts, which can collectively be called "graphics." Electronic publishing systems try to make it easy to create and work with such graphics, and in most cases they do.

There is a wide variety of graphics programs that allow the creation of line drawings (such as Corel Draw or Adobe Illustrator) with much less effort than by hand. There are also specialized software programs for working with photographs (e.g., Adobe Photoshop, ArcSoft Photostudio; Figure 6.2) as well as creating animations (e.g., Macromind Director). Existing illustrations and photographs can be put in electronic form using a scanner. Digital cameras can be used to take photos that are then loaded directly into the computer. Large collections of electronic clip art and digital photos are also available (now provided on CD-ROMs), making it quick and inexpensive to find graphics for any type of project or content. Finally, many other types of software, such as spreadsheets (e.g., the popular Lotus 1-2-3) or Computer Aided Design (CAD), programs generate graphics based on input data, which can then be integrated into other documents using page layout or word processing software.

Although there are many ways to create and manipulate graphics information in electronic form, creating attractive and effective graphics (whether in electronic format or not), however, still requires artistic and graphic design

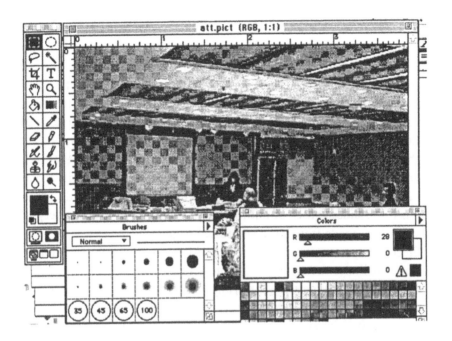

Figure 6.2   Photoshop (Adobe), a Popular Image-Editing Program

skills. Even though there is a plethora of graphics creation software and hardware, the expertise needed to use these tools well is almost always in short supply. On the other hand, graphics technology allows a little bit of ability or training to go much farther — since technology amplifies human skills and knowledge. The biggest obstacle to taking advantage of graphics seems to be the difficulty that most people have in visualizing information in graphical forms. There are also principles and techniques to be followed when creating visual presentations (Moore and Dwyer; Pettersson; Tufte).

Display size presents a limitation for certain kinds of graphics. With very large schematics (e.g., blueprints, wiring diagrams, maps), it is difficult to show enough detail or as much of the graphic as desired. On the other hand, the viewing program can allow scrolling in all directions (effectively producing a virtual screen size) or can provide zooming capability that allows progressively more detail to be revealed. Considering that large schematics are a problem in print form as well, electronic forms may actually be more usable and less costly to produce. Plus, it may be possible to present additional viewing options such as three-dimensional perspectives or coloration/shading. However, in many cases, such graphics need to be designed specifically for electronic display from the raw data used to generate the schematics.

Electronic Presentations

One category of electronic publications that is especially relevant to workplace learning is presentation (a.k.a. slide show) programs and hardware. Historically, the materials used for a presentation were overhead transparencies or 35mm slides. Overhead transparencies represented a tremendous advancement in visual aids — they are quick and easy to create and allow writing (or annotation) using a marker pen during the presentation. Use of overheads usually results in a more effective presentation because they require some degree of advance planning and organization (unless they are made up on the spot). However, overheads can also result in boring and dull presentations if the slides consist of nothing more than lines of text in a very small type size (i.e., made from copying printed pages).

Electronic presentation software (e.g., Microsoft Powerpoint, Adobe Persuasion) facilitates the creation of more visually interesting slides. It is easy to choose different type styles/sizes, apply color to text or backgrounds, add graphics, and employ special effects (such as wipes and patterns). Graphs or charts generated by spreadsheet or mathematical programs can be easily pasted into slides. Multimedia elements (e.g., audio, video, animation) can also be added. To the extent that the information to be presented is already available in electronic form, it makes it that much easier to create electronic slide shows.

Electronic presentations require special hardware to deliver in the form of a) LCD (Liquid Crystal Diode) projectors that project the computer display to a wall or screen, and b) LCD tablets used in conjunction with a standard overhead projector. Such devices are expensive and not always easy to find; thus, delivering an electronic presentation may not be a viable option in certain circumstances. Once a presentation has been created using slide show software, however, it can always be printed out on overhead transparencies and given in a traditional manner. In addition, almost all presentation programs allow multiple slides (in reduced size) to be assembled on pages to be used for printed handouts. It is also possible to distribute electronic presentations in disk or CD-ROM form, or to load them onto a network for people to view or obtain copies. The virtue of this latter approach is that you can take advantage of the interactive capabilities of an electronic presentation without the need for special presentation equipment — and each viewer gets his or her own personal showing.

Toward a Paperless Workplace?

Any discussion about electronic publishing raises the obvious question: why bother printing documents anyway? If they are going to be created, assembled, and distributed in electronic format, why not just "consume" them that way as well?

This question brings us to some of the supreme advantages of paper as an information delivery medium. It is highly self-contained (i.e., needs no power outlets or batteries), extremely reliable (printed documents almost never fail to work), and is very flexible in terms of where and how it can be used. There is also the issue of resolution and aesthetics. Until recently computer displays could not compare to the printed page in terms of the amount of detail that could be presented per square centimeter. And, for a long time, paper has been a very cheap material, which made the economics of printed documents very favorable.

All of these advantages, however, are slowly beginning to change. While paper is still relatively inexpensive, distribution costs (i.e., mail, shipping) for printed materials are increasing. Electronic delivery of documents saves the cost of paper and shipping. More important in many cases, delivery time is nearly instantaneous. As a greater percentage of individuals have access to a computer in their workplaces and homes (not to mention laptops and handheld machines), the issue of easy accessibility becomes less of a problem. Finally, screen display capabilities continue to improve with each successive generation of computer systems to the point where it will soon be possible to create screens with the same kind of detail found in print materials.

Furthermore, electronic media can offer so many more features than printed materials. An electronic document can have multimedia components (see Chapter 9). They can also include hypertext features which allow "hot links" among items. The world wide web (discussed in Chapter 14) illustrates how hypertext can be implemented in a global network environment where the links in a document are connected to other documents located on machines distributed around the world. Another critical feature of electronic documents is that they allow searching — something that is vitally important when large databases of information are involved. Finally, electronic documents require no physical space at the user's site for storage — an important consideration in crowded work environments (e.g., a factory site, space shuttle cabin, home office, car, or a typical office cubicle).

Desktop Publishing at the Economic Development Institute of the World Bank

The World Bank is a huge, global organization with many departments and offices around the world, many of which carry on educational or training activities, and produce an enormous number of publications and reports in multiple languages.

The Economic Development Institute (EDI), which is specifically tasked with conducting learning programs for the World Bank, produces a great deal of material directly related to instruction and learning. These documents include a variety of case studies and course materials to be used in teaching activities, a publications catalog, and a regular newsletter. Many of these documents are produced using desktop publishing technology of the kind described in this chapter. Documents are written by staff members or consultants using word processing programs, transmitted electronically over networks to the publication group in EDI, edited and composed into final form using page layout software, and then sent to the Bank's printshop (or an outside vendor) for final production. Since the content of these documents is economic in nature, they often involve an extensive number of charts, tables, and graphs, which pose special challenges for formatting and data transfer across different programs (e.g., spreadsheets, databases, statistical analysis).

John Didier, publications officer for EDI, describes the impact of desktop publishing on their publishing activities: "In certain ways it increased our publishing because it allowed us to improve the presentation of materials and gave us more control over the formatting, making training materials more accessible and easier to read. Quality of presentation has improved ... but cost-effectiveness is hard to measure. Turnaround has improved as well, but in some cases it may in fact increase the number of revision cycles — you tend to revise more because you have more control over the 'proofs' and you can fine tune right up to the end, and hopefully produce a better product."

Like all modern organizations, the World Bank is exploring the use of new technologies for information dissemination. For example, some documents (such as the newsletter and training schedules) are being made available on the web, and CD-ROM is also being investigated. For example, EDI has developed a prototype version of its publications catalog on CD-ROM and is testing at field locations in different countries. Since the readership for EDI and World Bank publications is often located in developing countries with limited access to computer technology (and networks), they have to proceed cautiously in terms of using the web and CD-ROM as distribution methods.

Once these countries do get suitable equipment and network connections, they are immediately able to take advantage of the extensive information resources that an organization like EDI and the World Bank can provide. Edith Pena, another publications manager for EDI, comments: "I think CD-ROM is pretty international and useful everywhere. Developing countries are very capable of handling CD-ROMs because they are new to technology and can buy the best as it comes out, whereas we're in some cases still using older machines."

One of the most interesting challenges for the World Bank and every other organization is how to make the transition from paper-based to electronic publishing using media such as CD-ROM and the web. Edith Pena explains: "CD-ROM is making us recreate the materials with the subject matter expert so that what is on the CD-ROM is an improved version of what existed before on paper. We want to add functionality and use the technology as much as possible to allow people to use the material well, which requires the redesign of the material and content." Didier adds: "The value-added of having something on the web is to make it broadly accessible and transmit large amounts of material, but at the same time make it accessible to different audiences. So what we might do is deliver a book-length manuscript but pull out or prepare summaries at different levels of detail, for example, the general reader, the high-level policy maker, and something for people who want to get into the technical nitty gritty. So, we have to think through how best to do that."

All of these factors are sufficiently compelling to cause many individuals and organizations to look seriously at publishing documents completely in electronic form — and the increasing number of web publishing efforts illustrates this trend. It is interesting to note that almost all the programs mentioned in this chapter now offer web formats as an output option. At present, however, most people are content to take advantage of electronic publishing tools to create documents in a quick and inexpensive manner, but still put them in printed form for distribution and use. Some organizations are distributing their publications in CD-ROM form, which is an interim strategy since it allows most of the advantages of electronic media while still using a physical medium. CD-ROMs (discussed in detail in Chapter 9) are useful for people who do not have easy access to networks. They also provide some degree of security and copyright protection for the information being distributed.

The Look and Feel of Learning

The final issue to ponder is how electronic publishing might alter the nature of learning and training. Some people feel that printed documents give the expression of ideas more permanence and concreteness. They lament that information presented in electronic format does not seem to be assimilated as well and is less likely to be remembered or reflected upon. Even when the information is easily retrievable, people may be less likely to review it.

But electronic publishing does make it easier for almost all individuals to produce high-quality documents, and, in this sense, it is an equalizer among people of different circumstances. For example, persons with disabilities can generate publications that are no different from others (perhaps with the help of adaptive software or hardware). Individuals in poor rural or urban areas, or in developing countries with limited resources, can still contribute to and produce publications of professional quality. Electronic publishing technology lowers barriers that might exist with traditional methods — unless, of course, you have no access to a computer and suitable software.

From an organizational perspective, electronic publishing allows information to be collected, assembled, and delivered faster and less expensively than in paper form. Electronic publishing should allow training materials to be produced more efficiently. Indeed, electronic publishing allows new forms of on-the-job learning such as performance support systems (see Chapter 13). Electronic publishing can also significantly affect the quality-control aspects of documents. Since it is possible to generate documents "on-demand," the master version can be continuously updated. Corrections and improvements from users can be made as they are reported, rather than on a periodic, sporadic basis.

There are also limitations relative to the effectiveness of electronic publishing: a) the need for everyone to be trained (and retrained) on how to use the hardware and software (a significant investment of time and resources), and b) the need for people involved to have some degree of text and visual design skills to prevent poor results. In essence, electronic publishing raises the threshold of literacy (both computer and media) for employees. It provides new opportunities for preparing learning programs and materials, but only for those with the appropriate skills.

Summary of Key Ideas about Electronic Publishing

- Improved ability to create professional-quality publications
- Document production time can be reduced significantly
- Publishing activities can be distributed locally to appropriate groups and individuals
- All employees involved in electronic publishing need some basic document design skills and knowledge of software
- Good writing/communication skills are still fundamental to electronic publishing efforts
- Use of on-line facilities (i.e., LANs, e-mail, groupware) streamline the publishing process
- Broader range of media and software skills needed when publications involve multimedia components (audio, video, animation)
- While print/paper distribution may still have some cost/benefit advantages for the present, electronic document delivery is the future

---

References

Carroll, J.M. (1990) *The Nurnberg Funnel.* Cambridge, MA: MIT Press.

Gropper, G.L. (1991) *Text Display: Analysis and Systematic Design.* Englewood Cliffs, NJ: Educational Technology Publications.

Horn, R.E. (1989) *Strategies for Developing High-Performance Documentation.* Waltham, MA: Information Mapping Inc.

Horn, R.E. (1992) *How High Can It Fly? Examining the Evidence on Information Mapping's Method of High Performance Communication.* Lexington, MA: The Lexington Institute.

Jonassen, D.H. (1988) *The Technology of Text: Principles for Structuring, Designing, and Displaying Text. Volumes I & II.* Englewood Cliffs, NJ: Educational Technology Publications.

Moore, D.M. and Dwyer, F.M. (1994) *Visual Literacy: A Spectrum of Visual Learning.* Englewood Cliffs, NJ: Educational Technology Publications.

Misanchuk, E.R. (1992) *Preparing Instructional Text: Document Design Using Desktop Publishing.* Englewood Cliffs, NJ: Educational Technology Publications.

Pettersson, R. (1993) *Visual Information* (2nd ed.). Englewood Cliffs, NJ: Educational Technology Publications.

Tufte, E.R. (1996) *Visual Explanations.* Cheshire, CT: Graphics Press.

7 | Television and Video

One of the most powerful and influential media of the 20th century is television. Clearly, television has profoundly affected our culture and the way we think and learn, as well as how we present and process information. It makes sense that television — and its offshoot, videotape — should also have a significant impact in the realm of education and training in the organization and workplace.

Why Television and Video Are So Important for Learning

Before discussing the use of television and video in formal training settings, we should acknowledge that commercial television (including public broadcasting) is responsible for a tremendous amount of informal or incidental learning (particularly with children). While most of what people learn from television is sociocultural in nature, there is a certain amount of information that has to do with technical subjects or business skills which may be useful to individuals in their work. While it is unresolved whether the overall influence of commercial television on society is positive or negative, there can be no argument that it exposes everyone to a great deal of information — much of which is retained and absorbed.

More important, television creates an expectation of how information should be presented, i.e., in a highly dynamic and multimedia form. Television relies on quickly changing images, drama, sound and music, comedy, and various other devices not often found in traditional instructional materials. Traditional training offerings, to put it bluntly, are deadly dull compared to most television programs. This realization was the impetus behind the work of the Children's Television Workshop, which created Sesame Street

Table 7.1   Programs Offered by PBS' The Business Channel in 1997

Japan/China Business series (live)
Leadership and Business Ethics (live)
Peter Senge on Competitive Advantage (live)
AMA's Annual Briefing for Secretaries (live)
Making Managers Into Leaders (live)
Management Skills & Issues series (pre-taped)
Competing in a Global Marketplace series (pre-taped)
Business & Technology series (pre-taped)
HRD/Communications series (pre-taped)
Innovative Leadership series (pre-taped)

and other television programs for children (Lesser). It also provides the basic rationale for educational television which has been used for decades in public schools and at the post-secondary level. PBS and many community colleges around the country produce and show thousands of hours of programming directly tied into courses (Annenberg/CPB; Zigerell). PBS also operates The Business Channel, which offers a series of workplace training broadcasts in live and pre-taped format (see Table 7.1).

Actually, there is another rationale for the use of television in education besides its unique presentation capabilities — the cost-effectiveness of broadcasting as a delivery method. Television programs can be viewed by an unlimited number of people within a broadcast area without affecting the delivery cost. Even when television programs are delivered via cable, satellite, or microwave transmission (which involve additional costs for each receiving site), the number of people viewing programs is only limited by the size of the television and the room. The economics of television makes it an ideal medium for a large audience distributed over a broad geographical area. Furthermore, videocassettes make it possible to selectively distribute programs, albeit without the cost advantages of broadcast delivery.

Television programs of commercial quality, however, are expensive to produce and require special resources, such as camera crews, technicians, engineers, directors, editors, script writers, and possibly actors (not to mention production facilities). Because the development of education and training materials is almost always done on a shoestring budget, most instructional television follows a different model of production from commercial television, often called the "candid classroom" approach. In this model, a traditional classroom lecture or presentation is videotaped, often illustrated with graphics (or in the worst case, writing on a whiteboard). It comes as no surprise that instructional television of the candid classroom type is no more effective than untelevised classroom instruction, since it hardly takes advantage of what television has to offer as a powerful communications medium. It still can achieve the economic benefits of television distribution though, and hence is justified on this basis.

The Technicalities of Television

One of the most intriguing aspects of television/video as a medium is that, while almost everyone is very familiar with it (having watched hundreds of hours every year), few people have any idea how it is created or the technical considerations associated with TV production and delivery. This probably should not be too surprising since the same can said about many other technologies in our daily lives: electricity, telephones, automobiles, etc. So when it comes to making good decisions about the development and use of television for education and training, it's not too difficult to understand why many mistakes are made — and why the medium is used so ineffectively.

While the programs seen on a television set all look the same, they can arrive there in many different ways (Figure 7.1). The classic form of television transmission is via VHF/UHF radio frequencies that are broadcast from a tower over a limited geographical range (e.g., 10 to 20 miles). This requires a broadcast license from the Federal Communications Commission (FCC). Today, most people obtain their TV programming via a cable TV (CATV) operator which involves transmission over a coaxial cable from the cable company's satellite receiving location to their home. CATV is also regulated by the FCC. It is also possible for an organization (or individual) to create a private cable TV service (closed-circuit TV or CCTV), provided that the cables stay on their own property and are not involved in commercial service. Of course, anyone can buy his own satellite dish and directly receive programs being transmitted via satellite frequencies. However, most satellite transmissions are encrypted, so they cannot be freely received without decoding. Finally, there are a number of direct broadcast satellite (DBS) TV providers who provide programming to sites that have digital satellite receivers and dishes that include decoders for the encrypted transmissions. Use of satellites for noncommercial transmission is not regulated by the FCC, which is one of the reasons why many organizations prefer this form of television delivery.

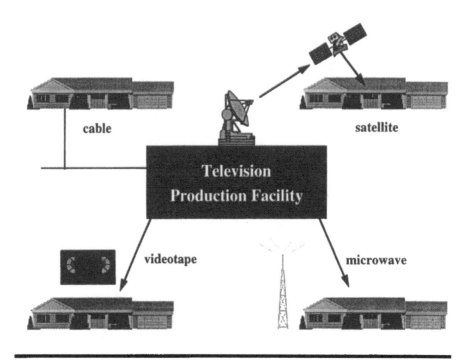

Figure 7.1   Different Forms of TV Transmission

In addition to all these commercially available TV delivery methods, there is one additional method unique to education: ITFS (Instructional Television Fixed Service). This involves microwave transmissions that are received by special antennas tuned to ITFS frequencies. Microwave transmission requires line-of-sight and has a range of about 10 to 20 miles. ITFS was designed for local instructional television broadcasting and has been used for many years by schools and colleges to deliver classes in their own region. For example, Stanford University began using ITFS to deliver engineering courses to local companies in 1969. While ITFS transmission is limited in range, it can be delivered to a wider geographical area by being routed through cable TV or satellite systems. That illustrates an important point about television programming: once it leaves the TV switcher, it can be transmitted in any number of ways depending on the equipment and resources available.

Videotapes

Videotape is the primary means for physically storing and producing television programs. Most television programs today are recorded on videotape for broadcast. There is relatively little "live" broadcasting (except for news and sports shows) because it is inherently unreliable and limits the kind of TV production possible. Even "live" programming tends to rely heavily on "roll-ins," which are prerecorded segments on videotape. Programs that are recorded can be edited to fix mistakes, add music or sound, mix together different footage, and employ a variety of special effects. While most of these things (except fixing mistakes) can be done in real-time, editing videotape affords a lot more precision and much less stress on production staff.

But even more important is the role that videotape has taken in delivering television programs. A large proportion of educational programming is now provided in videotape rather than broadcast form. Videotape delivery is preferred when the viewing audience wants the convenience of watching programs at their convenience rather than at a scheduled time. Since this is true for most employees and adult education settings, videotape has come to be the default for training and post-secondary learning. Furthermore, videotape delivery may be more cost-effective than other forms of television broadcasting when the audience is relatively sparse and dispersed over a broad area (e.g., a typical organizational training scenario). While such an audience can be reached via satellite transmissions, this is probably not going to be worth the cost of using a satellite if there is a small number of people involved (e.g., less than 100). Furthermore, there are time zone considerations to deal with when the geographic area gets larger — toward being nationwide or international.

Making Video Learning Programs for the Workplace

There is a variety of different methods and approaches to creating television and video (Zettl, 1992, 1995). The single most important element in making video is the camera. Since a large percentage of the population own and have used camcorders, there is little mystery as to how they work. However, the VHS camcorders used for recording home video and the sophisticated Beta-cam or Hi-8 camcorders used for professional video production are as similar as a standard car is to one outfitted for racing — they both have the same basic functionality, but the latter delivers a lot more performance! Professional-level cameras have better quality video and sound recording capabilities (due mostly to the Betacam or Hi-8 formats), and they are designed to work with external microphones and lights — two aspects vital to high-quality video. Video can be shot on location or in a studio. In a studio, there usually are multiple cameras hooked into a switcher, which allows a director to shift from one camera to another during the recording process. When video is shot on location, only one or two cameras are likely to be used, and the footage is mixed together during the editing phase.

Editing is the second most important element of making a video. It is the editing process that puts the "raw" footage into final form along with adding sound, music, graphics, animation, and special effects. During editing, segments from the original footage are rearranged into different sequences along with manipulation of the sound track to create a certain "feel" or message. This is a highly creative process in which the resulting program can take many possible forms, depending on the skills and talent of the editor and the camera work reflected in the original footage. On the other hand, the resulting program can be a more or less linear version of what was shot in the studio or on location, if that is desired. Most educational video tends to fall into this category compared to commercial programming, which tends to involve extensive editing.

New Television Technology

An important new development is occurring with television technology that is likely to make the whole video production process less costly and time-consuming — most aspects of video/television production are going digital, i.e., becoming computer-based. We have mentioned in earlier chapters about digitizing video footage using a personal computer. The camcorder (or a tape deck) is connected by a cable to the computer, and digitizing software transfers selected video sequences into computer files in a compressed format (such as Quicktime or MPEG). These files can then be edited on the computer and incorporated into programs for delivery via CD-ROM, networks, or the web. Digital video requires a lot of disk space — approximately 1 megabyte for every few seconds — so you need a very large-capacity hard disk (with many gigabytes) to store and work with it.

The significance of digital editing is that far fewer resources (equipment, facilities, and human) and much less time is needed to do video production. Indeed a single person can shoot video with a camcorder, then digitize and edit the footage in a fraction of the time it would take using traditional equipment — all with a personal computer system that costs only a few thousands of dollars (a fraction of the cost of traditional equipment). Furthermore, many digital editing systems actually offer additional functions and capabilities that are not available with traditional equipment (unless it is very expensive). If the person or team working with the digital system (and shooting the video) is experienced and talented, they can easily produce high-quality work that rivals that done with traditional production equipment and facilities, and they can do so much faster and for considerably less money.

Whether this development will translate into better quality educational programs is an open question. It does mean that trainers can make video within more limited timeframes and budgets than has been possible historically. There are, however, a number of other instructional issues that affect the way television has been used in training.

How Television Is Used for Training

There are four broad categories of television programming in the training domain: 1) technical skills, 2) product information, 3) sales/marketing, and 4) management. *Technical skills* can include how to operate or repair equipment, fundamentals or theory of a given domain (e.g., digital electronics, programming), and basic or advanced procedures for a specific job or task (e.g., installing fiber optic connections, risk management). In most cases, there is a need to demonstrate how things work or use graphics that convey concepts and relationships. *Product information* usually focuses on the features and functions of newly announced products, with demonstrations and discussions regarding the intended customer. *Sales and marketing* programs tend to focus on strategies and techniques with testimonials and strong motivational elements. *Management* programs are usually about leadership or motivation skills, as well as social/legal issues such as discrimination and rights. Of course, a program can be a combination of the different types, and most organizations who use television would likely produce programs of all three kinds for any major product line or job domain.

Television is also used extensively for continuing education in certain professions and occupational areas. For example, NTU (National Technological University) based in Colorado has offered graduate courses in engineering via television for many years, and a number of schools of engineering at major universities around the country also offer courses in television format.

Similarly, many universities offer degree courses in health-related subjects for continuing medical education. A number of professional organizations, such as the American Banking Association, the American Law Association, and the American Management Association, also offer televised courses to their members. Such courses tend to focus on new topics and "hot" issues of interest to a specific group of professionals.

National Technological University: Engineering Education by TV

The National Technological University (NTU), headquartered in Colorado, provides an interesting example of the large-scale use of television teaching for adult education. NTU, which is a fully accredited university, offers masters degrees in 10 engineering fields as well as a broad range of engineering-related continuing education courses. All courses are televised and delivered via satellite to small groups of engineers in their workplaces. The courses are developed and conducted by faculty from major universities all over the country. Courses are paid for by the corporations on a subscription basis, rather than by the individual employees. The faculty who teach the courses are paid an extra stipend by NTU on a per student basis.

NTU is not responsible for the development or quality of the programs; this is left up to the faculty and their institution. Consequently, most of the programming is strictly "candid classroom," i.e., videotaped versions of the professor's on-campus classes. Even though this makes for terrible television, the faculty who give the courses are often the leading experts in the country on the topics covered, making them worth listening to. Furthermore, the courses are usually being provided "free" to the participants, and the content is highly relevant to their work and careers. Under these circumstances, very boring TV programs can still be quite effective.

NTU is also not responsible for the management of classes locally. The subscribing companies promote the courses, decide which employees will be able to participate, and arrange on-site facilities for watching the programs. They also take care of operating and maintaining the downlink facilities and equipment (i.e., satellite dish, receiver, television sets). NTU courses are integrated into the companies' training offerings and professional development opportunities and may be supplemented by their own courses of a more proprietary nature, also delivered via satellite television using the same facilities.

NTU is a very successful model of technology-based training, having delivered thousands of courses and granted hundreds of masters degrees in engineering. By letting other universities take responsibility for the development of courses and having companies take care of managing course participation, NTU avoids the complications and risks associated with these two aspects of being an educational provider. Instead, NTU can focus on the logistics of delivering courses via satellite and coordination of the whole enterprise. (For more information, see http://www.ntu.edu)

A typical "show" in a continuing education program involves brief statements by a number of leading experts about a topic followed by a discussion among them about differences of opinion and interpretations of research results/application outcomes. There are also likely to be "roll-in" segments that feature demonstrations or specific case studies done on location. This format differs from the candid classroom model in that it involves conversational interaction among people (rather than a lecture) and it presents some real examples for viewers to study, all of which makes for more interesting television programs (although still quite pedantic relative to commercial television offerings). Much continuing education today takes the form of satellite teleconferences (see next chapter) which adds two-way interaction with the audience and presenters.

Business Television

Many large corporations and agencies have created their own private television networks for internal communications and, in some cases, customer education. These private networks are often called "Business Television" (BTV). Irwin describes over 80 BTV networks in the U.S., some of which reach thousands of sites and millions of employees. IBM, for example, operates a network called Corporate Educational Network (CENET) for management training which broadcasts multiple programs simultaneously to hundreds of sites all over the U.S. and the world. Other high-tech companies such as Hewlett Packard, Kodak, Tandem, and Motorola also operate their own networks for internal training purposes. Finally, the Department of Defense uses instructional television extensively in all branches of service and may well be the largest single producer of educational and training programs in the world.

Unfortunately, many programming formats that are highly effective and commonly used in commercial broadcasting are seldom seen in instructional television. For example, game shows are a staple format of commercial television that are rarely used in educational settings. Yet game shows are very motivational and quite effective in conveying information. Consider a "Wheel of Fortune" or "Jeopardy" type of program for conveying product information to employees, complete with prizes and employee players. Similarly, dramatizations such as soap operas or sitcoms could be very good ways to convey "soft skills" such as management strategies, policy issues, and organizational culture. For example, the World Bank employed a soap opera format in a television series for eastern bloc countries intended to introduce the basic concepts of free enterprise and market-driven economies. Indeed, there are many techniques that can be used to increase the effectiveness of television for learning, but they are rarely used (Cyrs and Smith).

The biggest mistake made in educational television relates to poor understanding and application of segment length. Because of the intense nature of television viewing, a program needs to be broken down into brief segments of 4 or 5 minutes long in order to maintain the viewer's attention and mental stimulation. Most commercial television programs handle this by constantly shifting from one scene to another, often as frequently as every 20 to 30 seconds. (MTV and music videos are even more fast-paced). Plus there are commercial breaks every 10 to 14 minutes. However, instructional television programs often go on, without any significant change of scene, for periods of 20 to 30 minutes. Even the most compelling content and the most effective speaker cannot hope to maintain the viewer's interest over such a long duration! Of course, breaking a program down into many small segments takes a lot of design and production work, elements that are usually lacking in instructional TV.

Interactive Television

One way to make television more interesting and effective for learning is to add an interactive capability via a student response system. Students have small keypads that allow them to respond to questions posed by the instructor. The responses are transmitted over a telephone line and are displayed on a personal computer at the instructor's location. This allows the instructor to continuously evaluate how well the information is being understood, and it makes learning active instead of passive. However, use of a student response system requires the availability of the keypads at the learning sites and a personal computer in the TV studio. Also this only works for "live" broadcasts. Many organizations, such as Hewlett Packard, Ford Motor Co., AT&T, Microsoft, the FAA, the Red Cross, and JCPenney (see following section) have used the One Touch student response system as part of their instructional television efforts.

Use of Interactive Television
for Training at JCPenney

The national department store chain JCPenney has used business television and videotape to distribute information for many years. In 1996, however, it radically changed the way associates were trained. JCPenney made a strategic decision to close its corporate university and to discontinue the practice of having managers travel to the company headquarters in Plano, Texas, for training. It also chose to stop the printing, warehousing, and mailing of paper-based training materials. These expenses saved were used to convert to distance-learning technologies, so that by 1997 more than 1100 stores had the ability to interact with instructors via interactive keypads. The new delivery system (called the Learning Place) included the use of the One Touch response units to allow two-way interaction. Each store has a television set in its training room and a number of One Touch response units, along with the necessary satellite downlink and reception equipment. Small groups of employees can watch a program and use the One Touch units to identify themselves and respond to questions in real-time.

According to Deborah Masten, Manager of HRD, the interactive television system has proven to be very effective and is a remarkable change agent. She cites a specific promotional campaign in which the stores that received their training via interactive television have shown sales gains almost twice as high as stores that only used printed sales materials. Furthermore, the interactive television system is viewed as a strategic business element by other groups within the company who want to use it to get their message out to employees. The interactive television medium, Masten notes, is "just-in-time, just down the hall, and just enough."

In the course of a year, JCPenney delivers hundreds of programs to stores from a single studio located in Plano. Programs typically range from 15 minutes to a couple of hours in length. The cost to design and deliver programs is kept low through the use of an automated studio that allows the instructor to control inputs (e.g., cameras, computer slides, video clips, etc.). A team of nine experienced instructors works with context experts from different departments in the production of programs. They may also work with other members of the HRD department if there are CBT, multimedia, or Internet components to a program.

One of the more interesting aspects of the Learning Place system is its extracurricular use. The system is used after hours for personal development classes in areas such as English language proficiency or SAT preparation. The classes are free to employees and delivered by outside specialists. JCPenney's interactive television program illustrates how a distance-training delivery system can be used quite broadly for employee education beyond specific job skills and knowledge.

Note: To hear more about the use of interactive television at JCPenney, see the interview with Deborah Masten at http://www.gwu.edu/etl.cases. html.

Emerging Developments: Stay Tuned

Thus far we have explored the use of television and video in their conventional forms. However, there is a number of emerging developments that are likely to change the role of television significantly, the most important of which is the convergence of television, telecommunications, and computing domains (Gross; Lochte; Luther; Negroponte). For example, television receivers on circuit boards can be installed in most personal computers today, turning them into television sets that can receive any form of TV transmission. This effectively eliminates the line between television and computer technology and allows television programs to be displayed along with computer programs on the same screen. Presumably when digital televisions come into use, the two will be able to interact and there really will be no distinction between TV and computer information.

More fundamentally, television principles are becoming as critical to the development of computer software as traditional programming concepts. Video sequences (in digital format) are an increasingly common and central element of interactive multimedia programs (discussed in Chapter 9). So all the production techniques and technical knowledge involved in the creation of television programs also apply to the development of video in multimedia efforts. Even though the equipment and methods for creating digital video differ, the rules about how to make good quality video and interesting programming are still valid.

This point is even more true in the domain of computer networks and the web (see Chapter 14). In the next chapter, we shall discuss desktop videoconferencing, which involves the use of computers to conduct two-way (interactive) video transmissions. While a videoconference is not a television program, many of the same elements are present, and the same production techniques underlie both. As desktop videoconferencing becomes more and more common, the principles of television production will become increasingly important knowledge for those involved in this technology. Furthermore, everyone involved in videoconferences (including users) will be exposed to the elements of video/television including cameras, microphones, lighting, color balance, frame and transmission speeds, etc. The irony here is that computing technology may ultimately be responsible for greater "TV literacy" than traditional television.

The implications of these developments for the educational domain are clear-cut: television and video are going to play an increasing role in learning in the context of interactive multimedia programs (especially those delivered via networks/the web) and videoconferencing. Even though instructional television has a long history, it has seldom lived up to its potential or matched the impact of commercial programming. But this will likely change in the future as television and video production becomes more accessible and affordable to all through the digital revolution.

Summary of Key Ideas about Television/Video

- Television shapes our culture and views and is a very powerful medium for communicating ideas.
- Television is a very cost-effective way to reach large audiences.
- Good television is expensive to produce; most instructional television is of the "candid classroom" format and not particularly effective.
- There are many ways to deliver TV programs, including: broadcast, CCTV/ITFS, cable, satellite, and videotape.
- Some of the most important aspects of TV production are type of camera used, lighting/sound, shooting location, and editing facilities.
- The costs/complexity of TV production are being reduced by the new generation of digital editing systems.
- In the training domain, instructional television has been used primarily for technical skills, product knowledge, sales/marketing, and management.
- Many corporations and industry sectors operate their own satellite TV networks for training, as do professional associations.
- The biggest single mistake in producing good instructional television, is not breaking content into short segments and not using a variety of formats.
- Instructional television programs can be made more interesting (and interactive) through the use of student response keypads.
- As computer and television technologies merge in the digital age, video is likely to become more commonplace in the training world and will likely be delivered in the context of multimedia programs on PCs.

References

Annenberg/CPB (1988). *Teaching Telecourses: Opportunities and Options, A Faculty Handbook*. Washington, D.C.: The Annenberg/CPB Project.

Cyrs, T. and Smith, F. (1990) *Teleclass Teaching: A Resource Guide.* Las Cruces. NM: New Mexico State University.

Gross, L. S. (1990) *The New Television Technologies* (3rd ed.). Dubuque, IA: W.C. Brown.

Irwin, S. (1992) *The Business Television Directory.* Washington, D.C.: Irwin Communications.

Lesser, G. (1978) *Children and Television: Lessons from Sesame Street.* New York: McGraw-Hill.

Lochte, R. (1992) *Interactive Television and Instruction.* Englewood Cliffs, NJ: Educational Technology Publications.

Luther, A. (1995) *Using Digital Video.* New York: Academic Press.

Negroponte, N. (1995) *Being Digital.* New York: Alfred A. Knopf.

Ziggerell, J. (1991) *The Uses of Television in American Higher Education.* New York: Praeger.

Zettl, H. (1992) *Television Production Handbook* (5th ed.). Belmont, CA: Wadsworth.

Zettl, H. (1995) *Video Basics.* Belmont, CA: Wadsworth.

8 Teleconferencing: Audio, Video, Computer, and Desktop

Teleconferencing is one of the most powerful forms of technology because of its capacity to allow direct interactions with and among instructors, experts, and students. There is a number of different types of teleconferencing, but they all provide some form of two-way interaction. In this chapter we shall examine audio, video, and computer conferencing, including the latest development — desktop videoconferencing.

Audioconferencing: Simple and Cheap

The most elementary form of teleconferencing is the audioconference which involves interaction via the telephone. In a formal setting, an instructor (either alone or with a class present) communicates with one or more remote sites, each of which has a small group of learners using speaker phones. Informally, one or more individuals will conduct a conference call with other individuals at different locations. While it is theoretically possible to connect an unlimited number of individuals and locations in an audioconference, in practice it works best for 5 to 10 people at no more than 4 to 5 sites. Once the number of individuals and locations becomes too large, it is difficult for all participants to interact easily.

Since telephones are universally available (at least in developed nations), access to the equipment is not usually a problem. Speaker phones, which are needed when a small group is participating at a single site, are relatively inexpensive ($200 to $1000 each). Long distance charges are likely to be involved, but these costs are also relatively modest (typically, $10 to $40 per hour depending on the distance and the service used). To connect multiple sites, it is necessary to have an audio bridge (a piece of telephone switching equipment), but this capability can be obtained through the services of a telecommunications vendor (e.g., almost any large telephone company) for a per person/per minute charge (typically about 20 to 40 cents/person/minute). The bridging charge may or may not include long distance charges — although it does when a toll-free (800) number is used. So a one-hour audioconference that hooks up a total of 25 people at 3 to 4 locations within the U.S. might cost $300 to $600 in bridging charges (plus the toll costs of the calls, if not included).

In a well-organized audioconference, an agenda, the list of participants, and any other necessary materials will be mailed, faxed, or distributed electronically to everyone well in advance. When the audioconference begins, the moderator will have participants introduce themselves (or perhaps just do a roll call of the participants, with each acknowledging his or her presence) and review the agenda. The moderator then initiates and manages the discussion of each agenda item or topic. It is the moderator's role to keep the discussion focused (and civil) and to summarize a topic when its allotted time is up or no further progress is being made. The moderator is also responsible for ensuring that all participants get involved, which means drawing out individuals who are shy and subduing those who are overly active. Needless to say, the role of the moderator in any teleconference is critical to its overall success and effectiveness (see FORDSTAR example in Chapter 18). One reason why audioconferences are often not successful in educational settings is because many instructors have no experience or skill as moderators — quite a different behavior than traditional classroom presentations!

A number of organizations and post-secondary institutions use audioconferencing extensively. For example, the University of Wisconsin has operated a state-wide audioconferencing system since 1966 which provides a variety of audio-based courses, serving up to 32,000 students connected by 200 sites across Wisconsin. The primary use of the system in Wisconsin is for continuing professional education, especially in the health-care field.

Some corporations use audioconferencing for sales and management training — a subject that seems well suited to verbal discussions and sharing of experiences by participants. Actually, audioconferencing can be used for any content domain since visual elements can be provided in print or electronic form beforehand. It does not work well, however, for long discourses (i.e., lecturing), which become very boring to listen to. So instructors accustomed to conducting one-way presentations in their classrooms must change their teaching style and methods to be effective via audioconferencing.

Satellite Teleconferencing

The most common form of teleconferencing used today is satellite teleconferences which involve one-way television broadcasts and two-way audio (telephone) links. Participants watch the presentation on a television and can ask questions via a microphone or telephone handset. Usually a large number of sites is involved (ranging from 10 to hundreds) with at least 20 or more people at each location. Clearly a teleconference can involve many people (possibly thousands), all of whom theoretically can participate in discussions, although in practice only a small percentage of people at each site actually do. Interaction usually takes the form of questions or shared anecdotes. It is also possible to equip each location with a fax machine so questions and information can be sent in written/printed form. This is especially valuable in international events (because of language differences) as well as when technical information is involved (e.g., equations or computer programs).

Satellite teleconferences tend to be one-time events devoted to a specific topic or issue (e.g., new product announcements or the latest research findings). The format for a typical satellite teleconference involves short presentations (5 to 10 minutes each) by a panel of experts, followed by discussion among themselves and responses to questions from the audience. Videotaped roll-ins of demonstrations or examples may be included in the presentations as well as graphics (slides). In teleconferences used as part of formal training courses, the participants may be expected to orally present answers to problems or summaries of assignments. In more general sales or product introduction teleconferences, audience interaction is more likely to be restricted to questions and suggestions.

Regardless of the exact nature of the teleconference, a good moderator is required to keep the discussion focused and on schedule (although the role of moderator in this type of teleconference is less demanding than an audioconference where there are no visual clues to follow). But the comments made in the previous section about the importance of distributing materials beforehand, especially an agenda and background on the presenters and topics to be discussed, apply equally to satellite teleconferences. Equally important is the presence of a site coordinator who takes care of organizing, supervising, and trouble shooting all aspects of the local site, as well as being the primary liaison with the teleconference developers. In some cases the moderator-coordinator roles can be combined, although if the site has many participants, it is best to make these two separate jobs.

Most satellite teleconferences involve origination from a single site which is broadcast to many locations. It is possible to have multiple broadcast sites, but this makes production and delivery much more complicated and expensive. All of the originating sites need to transmit to a central production facility where the segments are coordinated and then broadcast to the satellite. Alternatively, the coordinating site could hand off to each originating site that directly broadcast to the satellite, but this takes much more precise engineering. This type of multiple origination is easier to do with digital videoconferencing (discussed in the next section) and so is more common in this form of teleconferencing. It is far simpler to videotape contributions from multiple locations and then include the taped segments in a single broadcast (though this does not allow for two-way interaction with those locations — which might be important).

A particularly effective teleconference technique is to have a local activity at each site that prepares participants for, or follows up on, the broadcast event. A common way to do this is to organize a local panel of experts (or senior managers) to discuss the same issues or topics covered by the main presenters, but in the local context. This allows for greater participation at each site (i.e., the audience interacts with the local panel) and makes the teleconference discussion more relevant to the immediate needs of the participants at each location. Someone at each location (e.g., the site coordinator who may also serve as the local panel moderator) can summarize their discussion during the teleconference broadcast, and this can be compared to the general presentation. This technique of having local involvement makes teleconferences an ideal mechanism for collecting different perspectives across an organization or group and synthesizing them into a common view. It is a good example of how technology can address learning at the group and organizational levels.

Cost-Effectiveness of Satellite Teleconferencing

Ironically, even though satellite teleconferences are the most common form of teleconferencing, they are also the most expensive to deliver. A considerable amount of television production and engineering effort is normally involved in a satellite teleconference, thereby requiring a significant budget. The actual cost of satellite time (in the range of $500 to $1000 per hour depending on when and where) is a relatively small component of the total cost. Each receiving site must have a satellite dish and receiver, as well as suitable audio-visual equipment (television set, audio connections). It is also important to have a site coordinator at each location who ensures that the equipment is working, supervises the facility, and handles local arrangements. The equipment and staff needed at each site can belong to the organization (i.e., if the sites are branch offices or plants) or they can be provided by a hotel or vendor. A number of large hotel chains have all the necessary equipment at most of their hotels and can provide this service for a reasonable fee. Alternatively, there are teleconferencing vendors who will provide all the equipment and staff needed at whatever locations desired on a contractual basis. Either way, the total budget for a 1- to 2-hour satellite teleconference is likely to be in the $25,000 to $100,000 range. Of course, if an organization conducts teleconferences regularly and has all the equipment and staff in place, the costs will be much lower than if this is a special effort that requires the use of vendors or hotel services. Furthermore, teleconferences that involve international locations tend to be more costly because they involve greater preparation and logistics.

On the other hand, satellite teleconferences can be cost-effective under the right circumstances. If there is a need to share information with a large number of geographically distributed people in a very short timeframe, a satellite teleconference is a good choice. For example, if a company is announcing a new product to its entire sales force of thousands who are located at dozens of different offices, it is very likely that a satellite teleconference will be the most effective way to do this. This would also be true with the need to announce new technical or management knowledge or major policy/procedure changes. What makes a satellite teleconference an especially good delivery method for imparting new information to a large percentage of employees or members is its capability for allowing for a reasonable degree of participation and hence feedback from the audience to the organization.

Digital Videoconferencing

Digital videoconferencing provides even more interactive capability because it involves two-way audio and video transmission between each site. Indeed, videoconferencing provides the closest thing we have to "being there" via technology. Furthermore, it is relatively inexpensive and easy to implement (although not as simple and cheap as audioconferences).

Digital videoconferencing involves the use of compressed digital transmissions. A codec (compression/decompression) is required to send and receive the transmissions, along with a video camera and audio capabilities. There are a number of companies (e.g., PictureTel, CLI, VTEL) that market integrated videoconferencing units (that include the codec, television, camera, microphone, and speakers) in a variety of sizes and styles ranging in price from $15,000 to $50,000. These units can be "roll-abouts," which are about the size of a small desk and can be moved around, or they can be built right into the walls of a room to create a videoconferencing "suite." Each site must have a videoconferencing unit in order to participate in a videoconference, as well as a suitable telephone connection. Other ancillary equipment can be attached to the basic unit, such as an image scanner for transmitting visuals or a VCR for recording conference session, and transmitting prerecorded video.

Unlike satellite teleconferencing, which can be used to broadcast to hundreds of locations, videoconferencing is usually used to connect two sites with a small group of 10 to 20 individuals at each location. It is possible to link multiple sites, but this is likely to require the use of an MCU (Multipoint Control Unit) to switch the lines, plus additional monitors. Some of the newer and more sophisticated videoconferencing systems are capable of handling multiple sites without an MCU and have monitors which can display many sites on the same screen. The size and quality of the monitors is an important factor that affects the number of sites that can participate fully in a videoconference, as well as the nature of the interaction.

Videoconferencing involves the use of digital transmission via telephone lines or digital satellite links. In fact, there are quite a large number of different ways to transmit videoconferences depending on the picture quality desired, which is determined by the speed and capacity of the circuit used. For example, T-1 circuits are capable of transmitting about 1.5 Megabits per second (Mbps); switched lines can handle 56 Kilobits per second (Kbps); and ISDN offers 128 Kbps. Multiple circuits can be combined (or divided up) to obtain the desired transmission capacity. The way the information is compressed also affects the transmission quality, although most videoconference vendors now use the same standard format (e.g., H.320 or px64). At the higher transmission speeds (e.g., 768 or 384 Kbps), the video transmission rate can be a full 30 frames/second, which is essentially the same picture quality as standard television or video. At lower speeds (such as 112 or 128 Kbps) the display rate is likely to be 12 to 15 frames per second, which produces a somewhat jittery picture.

While all of this discussion about videoconferencing equipment and transmission options may sound complicated, in practice it turns out to be fairly simple. Usually the videoconferencing vendor will take care of arranging for and connecting the appropriate telephone connection (you just pay the bill!). To initiate a videoconference with another site, you dial up their telephone numbers and press a button. In fact, many of the more recent videoconferencing units have a preprogrammed handset or are controlled through a personal computer that automatically takes care of establishing the connection from menu selections. All videoconferencing units provide controls to adjust the camera position and focus, as well as the audio level and any ancillary equipment.

Videoconferencing is now widely used by most large corporations and organizations for meetings as well as training. The major benefit is reduced travel time and costs — which can result in substantial savings and productivity gains. The initial costs of the equipment and the ongoing telecommunications costs are easily justified by reductions in travel budgets. Conducting meetings via videoconference also makes it possible to include additional participants whose presence would not be likely if travel was involved (e.g., senior management or support staff). This typically results in improved communication and a broader knowledge base for decision-making.

One domain where videoconferencing seems to be particularly promising is medicine and health care. Many medical schools have established "telemedicine" efforts that involve the use of videoconferencing systems to teach medical students or offer continuing education classes to health-care professionals. For example, the University of Minnesota delivers medical lectures to radiology residents at hospitals in the Minneapolis area via a videoconferencing network. On a larger scale, the Georgia Statewide Academic and Medical System (GAMS) is a videoconferencing network that connects approximately 50 different health-care sites in the state along with over 100 other institutions for medical education and telemedicine applications. Of particular importance in the medical context are the abilities to have high-quality visual interaction (e.g., for patient interviews or observation) including the viewing of still images (e.g., X-rays, slides, photographs).

Video Teletraining at the U.S. General Accounting Office

Adele Ewing, Guest Author

The GAO introduced video conferencing capability in 1992 to support timely communication of audit assignments among team members residing in different locations. Shortly thereafter, the Training Institute (TI) initiated a two-year pilot program using this technology to deliver training across multiple locations. This pilot provided TI the chance to learn and experience the requirements of delivering professional course offerings to GAO staff at a distance. Four courses were selected for redesign and pilot testing over a two-year period. Training design, technical problems, and production issues were dealt with on a course-by-course basis. At the end of the two year pilot, the Institute accepted teletraining as a mainstream training delivery strategy.

The pilot was a challenge for training staff. They had to work hard to meet existing training demands, while developing experience with this new format. However, they saw learning how to use this new way of supporting classroom instruction as the "wave of the future," and there was no lack of volunteers. An intensive learning and support program was implemented to teach the basics of designing and instructing for video teletraining. A Teletraining Design Team was made responsible for providing design, development, and production. Technical support was established, and a manual was written to assist staff engaged in this effort, located on the Internet at <http://www.gao.gov/special.pubs/ti95001.pdf>

At the end of the two-year pilot, the Training Institute had learned enough to include a distance-learning program using two-way video as a tool to deliver training and education. The lessons learned included:

1. In redesigning courses for the video teletraining format, plan to build in frequent interactions; use panels, interviews, question and answer periods, and video clips to avoid student's eyes from "glazing over."
2. In building competence, instructors need to practice, practice, practice. There is no substitute. An instructor who is familiar with the equipment, the production crew, the process, and the television

camera will be more relaxed and confident in the teletraining classroom.

3. In developing quality, designers and developers need to convert visuals to the television format. Graphs, charts, and illustrated word pictures keep attention focused and aid retention. In televised teaching, however, these should follow the television presentation format.

4. In delivering the product, training administrators should plan and coordinate with site locations. While a content specialist/class facilitator was not found to be necessary at each site, it is important to have a coordinator who will handle the administrative details of class delivery such as the room, materials, handouts, and video. Scheduling a class can be more flexible; time zone differences should be considered.

5. In implementing the program, develop administrative policies and systems. Course materials need to be shipped ahead of time and practice sessions should be scheduled. Multiple sites need to be coordinated and registration issues resolved.

6. In designing and delivering the courses, provide assistance to instructional designers and instructors. When using any new technology, people are more comfortable when there is knowledgeable staff to help in development and implementation. Instructional designers and production staff who understand interactive training can provide the support needed to assure a smoother course development and delivery.

Today, this additional capability has given Training Institute managers, designers, and instructors another option as they look for ways to meet training costs associated with travel. So, what's next? Already, TI instructional designers and support staff are working to combine the capabilities of video teletraining and asynchronous computer learning over the network to meet training needs. This is our next "challenge."

Adele Suchinsky Ewing, an Instructional Technologist with GAO's Training Institute, has worked with distance-learning applications for 20 years. She has implemented Learning Resource Centers and distributed training programs, instructional television, and internet training. Her e-mail address is: ewinga.sup@gao.gov

Computer Conferencing

There is a variety of different ways to use computer networks for conferencing. Historically computer conferencing was text only, but more recent forms include graphics, audio, and video. In fact, the more advanced computer conferencing systems involve two-way video (i.e., desktop videoconferencing).

Computer conferencing can provide either asynchronous (delayed) or synchronous (real-time) interaction. Asynchronous participants read and respond to messages whenever they sign on to the system. In synchronous interaction, participants send and receive messages with others on the system at that moment. Asynchronous conferencing is preferred by most, because it allows people to participate in a discussion according to their own schedule. It is also very useful for international exchanges which span many time zones. On the other hand, some people like the immediacy and spontaneity of synchronous conferencing. There are many computer conferencing systems available (of both types), and many of these now reside on the web.

Computer conferencing has been used for years in higher education (Eastmond; Hiltz; Kaye; Waggoner), primarily for graduate-level programs. Since discussion of ideas is a central component of most higher education courses (regardless of the subject domain), computer conferencing is a natural fit. Indeed, computer conferencing allows a great deal more interaction among the students than is normally possible in a traditional classroom. This fosters collaboration and team work — highly desirable aspects of learning that usually do not get enough attention in conventional instructional settings. On the other hand, when introduced into an organizational setting, the opportunities for collaboration and team work made possible by computer conferencing correspond to normal work patterns.

Audiographics

Audiographics is one specialized type of computer conferencing system (synchronous) specially designed for instructional purposes. An audiographics system uses a personal computer to transmit audio and graphic images simultaneously. Each student has his or her own personal computer (equipped with a microphone) running the audiographics software. Alternatively, a group of students may be located at a site using a single computer with a large monitor or projected image and a set of microphones. The instructor and student are able to converse freely via the audio connection and can all see the graphics screens, which are controlled by the instructor. In addition, the computers may be equipped with a scanner or graphics tablet that allows any participant to show an image or draw something to be seen by everyone else. (Note that the same effect could be achieved by combining an audioconference with a fax machine or electronic whiteboard.)

Audiographics systems have been used for a wide variety of technical training programs. For example, at the University of Wisconsin (a pioneer in the use of this type of technology), audiographics is used to teach courses on technical Japanese, engineering, as well as music and art. Audiographics is especially valuable when the subject matter involves material that is visual in nature, such as schematics, formulae, and special character sets. It does require that each student or site have a personal computer equipped with the audiographics software, but this is not a particularly difficult equipment requirement to satisfy today.

Desktop Video

The most recent development in computer conferencing systems is desktop videoconferencing. A desktop video system is a personal computer equipped with a codec interface card, a small video camera, and a microphone. In essence, desktop video is a digital videoconferencing system scaled down to fit inside a personal computer. It requires a high-speed data connection, the faster the better. Most systems connect through LANs to a T1, ISDN, or ATM link to an intranet or the Internet. The video is then transmitted over the network and anyone else on the network can receive it and respond. Unlike full-size videoconferencing systems that use telephone connections and hence are limited in the number of sites that can be linked, a desktop videoconference can involve an unlimited number of participants, depending on the capacity of the network. However, the problem with display size still applies to desktop video; there is only space on the screen to show so many sites, although a large computer monitor can typically accommodate up to 12 small video windows.

The current picture quality of desktop video systems tends to be inferior to full-size videoconferencing systems as a consequence of the slower speed connections and smaller display monitors associated with personal computers. However, this does not seem to bother users that much, especially when weighed against the convenience of being able to use the personal computer in their office or home to participate in videoconferences. The fact that the personal computer is being used also makes it easy to capture and save video or still images to disk as well as access relevant text or data files. It is quite possible to have the videoconference in one window on the screen while running a program or displaying a document in another. This capability to "multiprocess" information in the same display area is an important aspect of making technology conform to the typical need to relate multiple sources at the same time.

Many companies market desktop video systems, including AT&T (Vistium), Intel (ProShare), and IBM (Person to Person), as well as all of the videoconferencing vendors. The most widely used desktop video system at present is probably CUSeeMe which was originally developed at Cornell University and distributed free to educational institutions, and is now available commercially from White Pine Software. Thousands of university/college faculty and students have obtained their initial exposure to desktop video using this system. Many large organizations are currently experimenting with desktop video, but the technology is still too new to be used widely in operational contexts. It is reasonable to expect that this technology will become widely used for personal and work-related communication. One of the major benefits of desktop video is that there are no telephone connection charges (although the organization must pay the costs of network connections), which removes any financial disincentive to use the system routinely.

Prospects for Teletraining

The costs of travel and the lost productivity associated with training have always been a major issue in all organizations. Teleconferencing offers a solution to the perennial problem of how to offer training in a cost-effective manner. While not every training activity can be conducted at a distance, a large percentage can. Reaction of students and employees to teletraining is usually positive. While most people would prefer to have in-person contact, they are willing to accept teleconferencing as a substitute because it reduces the need to travel and be away from home and office. In fact, once people get used to interacting via technology, many find it just as acceptable as in-person meetings.

Of course, the effectiveness of teleconferencing for learning purposes depends on how well instructors use it. Like audioconferencing, the moderator plays a critical role in managing the discussion and participation. Instructors who are unable (or unwilling) to take on the moderator role are going to be unsuccessful with teleconferencing. To be effective, teleconferences require a different pedagogy than is typically used in classroom instruction, but not all teachers and trainers will be aware of this or know how to change their methods. So any organization or institution that intends to adopt teleconferencing as a primary means of delivering education needs to retrain its instructional staff in terms of the different skills required (Ostendorf; Parker; Portway and Lane).

On a larger scale, teleconferencing requires a change in perspective on the part of organizations, managers, and employees. There is a well-established habit of conducting face-to-face meetings that is difficult to break. The fact that a "virtual meeting" with participants from all over the country or world can be convened almost immediately (assuming the necessary equipment is available to all) should be, however, a compelling force to break old habits. Furthermore, because teleconferences are generally much less expensive to conduct than in-person meetings, there is a strong economic push for adopting this approach. Given these benefits, we can expect to see organizational and institutional cultures change slowly to make teleconferencing the standard mode of interaction among individuals for education as well as all other business purposes.

Summary of Key Ideas about Teleconferencing

- Audioconferences provide an easy and inexpensive form of teleconferencing for small numbers of participants (10 to 50).
- Providing an agenda as well as other reference materials to participants before the teleconference begins is essential.
- All teleconferences need a moderator who manages the interaction and keeps the discussion focused.
- Satellite teleconferences involve one-way television and a telephone link for interaction with participants; they are used to broadcast to many sites simultaneously.
- Each teleconference site needs a coordinator to handle all organizational and technical aspects.
- Effective satellite teleconferences usually involve a mix of centrally produced and local events/activities.
- Satellite teleconferences are expensive to develop and deliver but they can be cost-effective when the audience is large.
- Digital videoconferencing involves two-way, face-to-face interaction among participants at a small number of sites (2 to 4).

- Videoconferencing is much less expensive than satellite teleconferences and can significantly reduce travel costs.
- Computer conferencing allows people to interact either in real-time (synchronous) or delayed (asynchronous) formats.
- Computer conferencing is inexpensive and affords a lot of flexibility in how groups interact.
- Desktop videoconferencing allows PCs to be used as videoconferencing units, and networks are used to transmit the video.
- Teleconferencing is becoming an increasingly common means for conducting training and education activities.

References

Duran, J. and Sauer, C. (1996) *Mainstream Videoconferencing*. Reading, MA: Addison-Wesley.

Eastmond, D.V. (1995) *Alone But Together. Adult Distance Study Through Computer Conferencing*. Cresskill, NJ: Hampton Press.

Hiltz, S.R. (1994) *The Virtual Classroom: Learning Without Limits via Computer Networks*. Norwood, NJ: Ablex.

Kaye, A. (1992) *Collaborative Learning Through Computer Conferencing*. New York: Springer-Verlag.

Ostendorf, V. (1995) *The Two-Way Video Classroom*. Littleton, CO: Virginia Ostendorf Inc.

Parker, L.A. (1984) *Teleconferencing Resource Book*. New York: Elsevier.

Portway, P.S. and Lane, C. (1992) *Teleconferencing & Distance Learning*. San Ramon, CA: Applied Business Communications.

Schaphorst, R. (1996) *Videoconferencing and Videotelephony*. Boston: Artech House.

Waggoner, M.D. (1992) *Empowering Networks: Computer Conferencing in Education*. Englewood Cliffs, NJ: Educational Technology Press.

9 Interactive Multimedia — Extending Computer Based Training

Apart from the Internet and the web (which will be presented in Chapter 14), one of the most exciting and important developments for technology-based learning is the emergence of interactive multimedia capabilities. These developments enable the HRD field to significantly expand the reach and range of computer-based training (CBT). Indeed, these capabilities are fundamental to what can be accomplished on the web as well as what is possible in digital publications, video, and teleconferencing (discussed in the preceding three chapters).

Putting It All Together

What exactly does interactive multimedia involve and include? First, it refers to the capability to integrate information in digital form for any sensory modality and deliver it on a single computer system. At present this includes text, graphics (as defined in Chapter 6), sound, video, and animation. Generated speech output and speech recognition are also possible, although not very refined at present (recorded speech works fine). In the future, gestures and smell may be added to the list.

To understand why multimedia is so valuable for learning in the work force, it is important to remember that, historically, computers could only handle textual information. This new ability to be able to store and display other modalities is a powerful capability that has emerged only recently. In the past, a variety of different and completely incompatible technologies were needed to develop and deliver multimedia information, such as cameras for photographs and slides, tape recorders for audio, and camcorders/VCRs/televisions for video. With multimedia, all of these different forms of information can be developed in digital form using computer software and hardware and stored/presented using personal computer technology.

Multimedia technology has dramatically simplified the development process for all media and made it possible for anyone with the right hardware and software to develop multimedia presentations that can be run on inexpensive and commonly available personal computers. For example, anyone with a digital camera can have his or her photographs immediately available for display or inclusion in a program, simply by loading them into a personal computer from the camera. In addition, they can do a variety of modifications to the photos (e.g., resizing, cropping, shading, coloration) once they are loaded into the computer. The time and special skills required to develop film (not to mention photo processing) have been eliminated, making the entire process of using photos in computer materials much faster and easier. Similarly, the creation of sound, video, and animation sequences is also much streamlined relative to older analog methods.

But there is another critical element to interactive multimedia technology — the interactive aspect. In the multimedia context, interactivity refers to the capability of the program to respond to user input and in doing so provide unique experiences. The simplest form of interactivity is the selection of items from a menu or list of options. In so far as users decide what they want to see in a program, or when they are ready to go to another screen through their selections, they are interacting with the computer. Compare this with what happens when you watch television or listen to the radio/stereo. You can turn these systems on or off (and control basic functions like channels, volume, etc.), but you cannot actually interact with the content presented.

In most programs today, screens involve "hot links" (which are words/phrases or graphic icons which, when selected, connect you to new information — technically called hypertext). The connected information may be another screen of text (such as a glossary or more detailed explanation), an illustration/photograph, or an audio/video clip. Such links increase the interactivity of a program because it becomes possible for the user to follow many unique paths through an information database. We will explore the nature of such links in more detail when discussing web-based training in Chapter 14.

It is also possible for interactive sequences to become specific feedback messages to responses made by the user. For example, the program could present a set of questions and then provide feedback messages based on what answers the user provides. In certain kinds of programs called coaches or wizards, the program actually creates a profile of the user and can display suggestions based on the specific sequence of actions or responses made by a given user. Another kind of interactivity is the sequence of actions that occurs in a computer game or simulation — all determined by the responses of the user in a sequence of events and actions. This latter form of interactivity can be very extensive and fine-grained given the rapid speed and number of responses made in a typical game or simulation activity.

Interactive multimedia implies two very important capabilities: 1) to be able to present information in multiple modalities and 2) to allow the user to control the interaction to varying degrees depending on the nature of the program. To the extent that almost all forms of learning are enhanced by involving multiple modalities and by having interaction (responses and feedback), interactive multimedia has obvious significance to workplace learning.

Multimedia Technology: CD-ROM and Beyond

Today's personal computers have all the capabilities needed to run multimedia programs: lots of computer memory (RAM/ROM), fast processing speed, high-quality color displays, and sound output. This leaves just one additional issue — how to actually deliver a multimedia program to the user. Historically, programs have been distributed on floppy disks, which today have a capacity of about 1.4 megabytes (MB). However, when you realize that a typical multimedia program is going to require at least 400 to 500 MB (probably much more if it contains a lot of video), it quickly becomes clear that it cannot be distributed on floppies.

Enter the CD-ROM (Compact Disk-Read Only Memory). CD-ROMs have a storage capacity of about 600 MB. The exact capacity depends on the format used for recording. CD-ROM technology provides the obvious means for distributing multimedia programs; indeed, one could make the argument that CD-ROM made multimedia technology viable by providing a suitable delivery medium. Furthermore, CD-ROM is a very cost-effective delivery medium. When CD-ROMs are pressed in quantity, they become as cheap as a few cents each to produce. Considering how much information can be contained on a CD-ROM, it is clear why this technology is superior to paper from an economic perspective. Furthermore, CD-ROMs cannot be easily duplicated (like print or cassette tapes), so it provides a greater degree of copyright protection or security for proprietary information — a very important factor for many publishers and technical providers.

CD-ROM technology has the wonderful advantage of having basically one standard format (ISO 9660) that almost all programs and systems read, so there is a good chance that any given CD-ROM will work in any computer. And since all personal computers sold today come with a CD-ROM player as an option, people should have no trouble finding a machine that can read CD-ROMs. An equally nice development is the emergence of inexpensive CD-ROM recorders (CD-R drives) — priced at less than $500 — that make it relatively easy for individuals to create single copies of their own CD-ROMs at their personal computer workstation. All of the major authoring tools (see Chapter 11) support the creation of CD-ROMs, and the CD-R drives usually include special mastering software.

Despite the overall excellence of CD-ROM technology, there are some limitations. The first is amount of space — 650 MB is still not enough for a multimedia program with a lot of video, because video typically requires about 1 MB for every 4 to 5 seconds (even in compressed format). However, a new form of CD-ROM technology called DVD (Digital Video Disc) addresses this limitation and has about seven times the capacity (45 MB). An even more significant limitation exists because CD-ROM is a physical medium and requires physical delivery, so there are shipping costs associated with the use of CD-ROM, although much reduced from print. In addition, when materials are revised, it is necessary to send an updated CD-ROM to people and try to get them to replace the old one with the new one (sometimes a difficult task).

The solution to the delivery problem is to use on-line networks for the dissemination of interactive multimedia materials instead of CD-ROM. The web can be used to deliver any form of multimedia program to any computer anywhere. However, many machines do not have fast enough network connections at present to make real-time delivery of elaborate multimedia programs a realistic alternative to CD-ROMs. This is clearly a temporary limitation, and network delivery of interactive multimedia seems likely to become the mainstream form of dissemination, with CD-ROM being used for special circumstances where network access is not available or advisable.

Andersen Consulting:
Using Multimedia to Teach Business Practices

Andersen Consulting is a worldwide consulting firm with more than 40,000 employees in 47 countries. The firm's main training center in St. Charles, Illinois, delivers thousands of hours of classroom instruction, but the costs and productivity loss associated with travel for training are a significant organizational issue. Furthermore, there are growing questions about the instructional effectiveness of traditional classroom methods for teaching certain courses. These two motivations led the Professional Education Division to investigate the use of interactive multimedia for self-study, "point-of-need" learning.

The first course to be converted to this form was a business practices course that consisted of 65 hours of instructor-led training. Annual enrollment in the course was 3,000 employees, and the audience was expected to increase because of additional hiring. The course teaches basic business functions using a case study company; the learners' goal is to conduct a business review of the company's operations and make recommendations to improve the business and customer satisfaction. The resulting 40-hour multimedia program consists of 15 modules and is distributed on CD-ROM. It takes the form of a simulation and features video segments for briefings and interviews with employees as well as working models of all the key functions (e.g., sales, product development, cash management) in the company.

The value of the new business practices course is summarized by Acovelli and Nowakowski (1994, p27): "In a traditional classroom setting or in paper-based self-study, tasks are often isolated from one another and from a realistic context ....On the job, consultants should be able to understand how the people and processes of a client interrelate. Consultants should be able to see connections and determine effects as they are gathering information; it should be a seamless process. Effective learning systems, then, are those designed to make learning more valuable by simulating tasks that have the look and flow of those found in the real world ....Interactive multimedia computer learning systems can do this." In addition, it is estimated that the course saves Andersen Consulting an estimated $10.5 million annually in travel expenses.

> ➤ increased motivation
> ➤ appeal to different learning styles
> ➤ more realism
> ➤ facilitates multilingual presentations
> ➤ higher retention
> ➤ better comprehension
> ➤ improved transfer of skills

Figure 9.1   **Benefits of Multimedia for Learning**

Learning Benefits of Interactive Multimedia

To this point we have discussed the nature of the technology, but what exactly are the benefits of interactive multimedia materials for learners (Figure 9.1)?

Probably the single most important impact of multimedia materials is increased motivation. By virtue of involving more than one sensory modality and requiring user responses, interactive multimedia programs capture more attention and create greater engagement on the part of workers/learners. Since getting a person's attention and keeping them interested in the material is one of the most basic aspects of learning, this capability of multimedia is critical, especially in the context of self-study or distance-education settings where motivation to complete courses may be weak.

Another aspect of being able to present information in multiple modalities is that it accommodates a broader selection of individual learning preferences. It is well established that people vary in their cognitive styles, and some of these differences have to do with preferred sensory modes (e.g., aural vs. visual information). To the extent that an interactive multimedia program can present information in a wider range of modalities, it should be more appealing to a wider range of learners.

An additional potential benefit of multimedia is more realism. Photographs and/or audio/video clips make the content more concrete. A common complaint made by learners at all levels is that learning materials and activities are not sufficiently relevant or realistic. By making it easier to include audio-visual elements in computer-based materials, the chances are greater that the content will be more "authentic." Furthermore, when simulation and case study methodologies are used, the degree of realism of the program increases even more.

Multimedia also facilitates the creation of multilingual materials because text captions or audio tracks with alternate languages can be added relatively easily. It is also possible to have video segments with speakers of different nationalities explaining or presenting information — not only dealing with the language issue, but also addressing credibility for a given audience. In light of the increasingly global nature of organizations as well as the increased diversity of populations in most nations, the capability to develop multilingual learning materials cost-effectively is a significant advantage.

It should be emphasized that all the potential benefits just described depend upon the design of the multimedia program to be achieved. Interactive multimedia materials can be well designed and effective, or poorly designed and ineffective. If designed properly, multimedia programs can be motivating and realistic; but this is not an inherent characteristic of the technology.

Multimedia Learning Environments

Interactive multimedia programs can be applied in a number of different learning settings, reflecting the broad range and increasingly ubiquitous nature of computer applications (Figure 9.2). Let's consider four such settings: 1) individualized use, 2) group (classroom) use, 3) electronic performance support systems, and 4) public access (kiosks). As a general rule, all of the potential benefits of interactive multimedia just discussed in the previous section apply to each setting.

Individualized Use

The traditional form of technology use in education is individualized delivery in which learners interact with the program alone, usually in some kind of learning center environment — although it could be in their offices or homes. This results in all the benefits of individualized instruction: the learners establish their own pace and determine what content is to be presented (assuming that the program allows this control). Depending on what kind of assessment activities are built into the program, learners should be able to evaluate their progress through feedback. Since they control the pace, learners should be able to take as much (or as little) time as they need to understand material and practice.

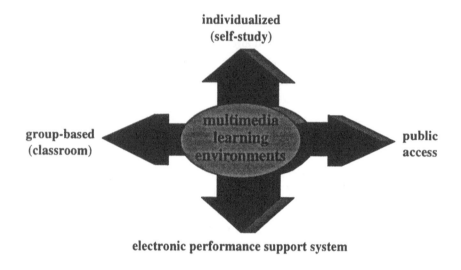

Figure 9.2   Multimedia Learning Environments

Group Use

While it is the traditional form of technology-based instruction, individualized use may not be the most important in the long run. Interactive multimedia is increasingly being used in group settings, especially conventional classrooms. By using multimedia presentations in a group context, it is possible to achieve the various learning benefits with many learners at once, which makes the whole effort more cost-effective from a hardware/software perspective. In this setting, an instructor or manager will use multimedia materials to deliver a more effective presentation and still be able to take advantage of the group dynamics and face-to-face interaction with the learners. At the very least, the presenter uses multimedia technology as a fancy audiovisual medium to show an electronic slideshow or web pages. Since most formal instruction today is still conducted in a classroom, it makes sense to take advantage of the benefits of multimedia in this setting.

There are a number of variations on the individualized and group settings. In many training centers, learners work in pairs or groups of three at a computer. The learning is still self-directed but a function of joint decisions and progress. There is every reason to believe that learning in pairs or small groups enhances the experience by virtue of the mutual support partners and team members provide each other. Another benefit is that having learners work in pairs or small groups means less equipment is needed. Similarly, learners can work together in pairs or small groups in a classroom setting as well, provided that the class involves activities or exercises for group interaction. Each pair or group can share a computer workstation, or they may be able to interact on-line if they are in a room that has a fully networked system.

Many universities and organizations have created electronic classrooms to take advantage of multimedia and computing technology. These classrooms feature some kind of large screen video projector (or LCD tablet) connected to a personal computer workstation for the instructor. This permits the instructor to run multimedia programs that the entire class can see and participate in. A further step is to equip each student with a workstation that is networked with others via a LAN (including the instructor's machine). This allows students to share information and data, and also lets the instructor send information to any or all students, or see individual screens. It's also possible for the instructor's workstation to be linked to networks or set up for digital videoconferencing, which provides for additional kinds of material or interaction from external sources.

Electronic Performance Support System

The electronic performance support system (EPSS) setting is quite different from the individualized or classroom environments (at least in intent). As we'll explore in more detail in Chapter 13, an EPSS is an approach to providing information or tools in the workplace by electronic means to help people do their jobs more effectively. In this context, interactive multimedia is used to show video demonstrations of how to perform a task, or provide a visual troubleshooting/installation guide with illustrations, photographs, and animations. While the contents of most EPSSs are strictly textual, there are many kinds of job aids that are much more effective with multimedia components. For example, any task that involves interacting with equipment is almost certain to be explained and understood better through the use of graphics or audiovisual sequences.

Kiosk

Finally, we should mention the public access (kiosk) setting. Many organizations have experimented with the electronic delivery of information or services in public locations such as shopping malls, building lobbies, libraries, museums, airports, hotels, information centers, grocery stores, etc. (Kearsley). The classic form of a public access system is the bank ATM, which first appeared in the 1970s and has been an enormously successful technology. In most cases, public access systems take the form of stand-alone kiosks and feature interactive multimedia programs. The multimedia capabilities are used to motivate people to use the kiosk or to present information (e.g., photographs or video segments of products/services). Because of the wide range of potential users for a kiosk in terms of age, educational and cultural backgrounds, or purpose, it is very difficult to design successful programs for this setting (multimedia or otherwise).

As this section has shown, there is a variety of learning environments, formal and informal, for which interactive multimedia can be applied. While the same set of potential learning benefits obtains in each, the different settings tend to emphasize certain characteristics of multimedia technology over others. This is important to remember when considering the design or costing of interactive multimedia technology; the nature of the learning context will determine the details of what particular aspects of the technology are most important and relevant.

Designing and Developing Multimedia Programs

While the focus of this book is not on the design and development aspects of technology, it is useful to have some general understanding of how interactive multimedia programs are created in order to manage or evaluate them. There are many books and web sites that provide detailed information on this topic (Blattner and Dannenberg; Hofstetter; Perry; Vaughan; also see Appendix B for web site suggestions). Indeed, any large bookstore will probably have dozens of up-to-date books on specific aspects of multimedia development.

Before any actual development work on a program begins, there is normally an extensive amount of analysis and design effort devoted to specifying the goals and purpose of the program, the intended audience and their characteristics, the typical environment that the program will be used in, as well as the anticipated system requirements needed to run the program. All of these steps, except the last one, are routine aspects of designing any program or curriculum materials and are really no different for interactive multimedia. The same methods and skills employed by a systems analyst or instructional designer for any project applies as far as the initial analysis and development of specifications are concerned.

Using Multimedia as Part of an Integrated Training Approach at Duracell

When Duracell opened a new battery manufacturing plant in Dongguan, China, to serve the growing Asian market, it decided to develop a comprehensive multimedia training program for the manufacturing processes involved. Duracell put together an international team to create and implement highly visual training materials which would rely on graphics, animation, digitized photos, video, and text in Chinese. The multimedia materials covered five modules: overview, health and safety, operations, quality, and fault finding. The modules included video footage filmed at various Duracell facilities throughout the world By using combinations of video sequences, animation, graphics, and photos, the program could provide a good understanding of the assembly process as well as the consequences of safety and quality considerations. Interactive exercises and questions dispersed throughout the modules ensured that trainees were learning the content.

While the use of multimedia was a key element of the overall program, it represented only one component of the total training package which included: manuals, classroom activities, printed job aids, actual equipment, and hands-on practice. The idea behind such an integrated training approach is that no single aspect of the training program should be expected to do everything or stand alone. The multimedia component ran on dedicated training computers (Pentiums with Chinese Windows) located in learning centers in the plant. This ensured that program performance was optimal and trainees were able to devote their full attention to learning when using the programs.

According to Neil Silverstein, a training manager involved in the project: "Our goal for this program was to create standardized training materials that would promote a safe environment enabling our work force to become productive as quickly as possible. The initial work force reaction was overwhelming. Members of the first team assigned to the computers fully embraced this new technology and the interactive programs. From the moment that the programs were installed, a crowd immediately formed around the workstations. Employees were anxious to get their hands on the touchscreen to begin their personalized training session."

Determining the system requirements is somewhat unique for multimedia projects because there is a number of considerations that affect design and development of the program. For example, the kind of display and sound capabilities, the typical amount of memory (RAM/ROM) available, the operating system, and the likely speed/type of network connection (if this is to be a network-based project) are all important factors in what multimedia capabilities can be used. On the other hand, certain assumptions can be made (e.g., the system will have a CD-ROM player, or a certain version of an operating system or web browser) which dictate the nature of the system that will be required in order to run the program. The latter approach makes for an easier development process (i.e., the characteristics of the run-time system are well specified), but this is not always a realistic situation if the program will be distributed in an open environment where there is no control over what type of system people have or use. In this circumstance, it may be necessary to develop multiple versions of the program that run under different types of system configurations.

Once the program has been defined, the design work can begin. In designing a multimedia program, there are two general sets of considerations to be taken into account: the user interface and specific multimedia features. The interactive capabilities of the program are a joint function of these two considerations. Some of the interactivity has to do with control options, while the rest derives from the content and learning strategies employed. There are numerous user interface issues to be addressed, such as navigation, error-handling, screen layout, response time, and so on. Most of these issues have been studied by researchers, and there are many guidelines available for designers to follow (Hix and Hartson; Howlett; Laurel; Shneiderman). Similarly, there are quite a few design choices that have to do with the multimedia features, beginning with exactly which media (i.e., graphics, audio, video, animation) are to be used. In most cases, decisions about multimedia features are largely determined by the content itself (e.g., illustrations vs. photographs) as well as the creativity of the designers. The capabilities of the authoring tools (see Chapter 11) as well as the system constraints also determine what multimedia features can be employed.

Once the design work has been completed, the development of the program begins (Figure 9.3). Actually, the transition from design to development is usually rather fluid because prototypes are often created to try out design ideas, and these are gradually extended to become the actual program. Plus,

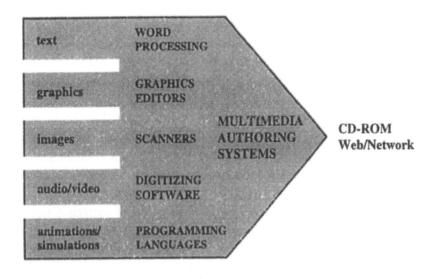

Figure 9.3   The Multimedia Development Process

development often reveals that design ideas don't work out as expected and need to be rethought. Development steps include writing text, creating graphics, taking or obtaining photographs, shooting video or recording audio segments, digitizing all these components, and programming any interactive sequences (such as input and feedback messages). This will require the use of various authoring tools and may also involve other software which is part of the content or task being trained (e.g., databases, CAD/CAM, proprietary systems).

Challenges and Opportunities

The development of interactive multimedia programs is an expensive and time-consuming activity, and this is a significant limitation to their wider use in organizations, especially those with tight budgets. Indeed there is a built-in irony to multimedia development — the more multimedia features employed and the more sophisticated the interaction, the more time and money are needed to develop the program. So there is a price to be paid for developing good multimedia programs, and it gets higher the better the program gets! Of course, the development costs can be reduced through the use of experienced developers, powerful authoring tools, and minimizing the amount of new material needed (e.g., use of graphics/photo libraries). But the real art to controlling costs and development time for multimedia programs is to be very selective in what multimedia features are used and the kind of interactivity provided.

Another ongoing challenge is the constant emergence of new multimedia capabilities. Even before a program is completed and fielded, it is likely that some of the multimedia features will be updated or enhanced in newer versions of software and hardware. While the program as developed may still work fine, it will seem dated or obsolete relative to the latest offerings demonstrated by vendors. In some cases new features are incompatible with older ones requiring that programs be revised in order to work properly. So multimedia development involves a continuous reworking of materials, not just to keep content current but also to change how multimedia elements are defined and programmed. This is especially true with the web, where the formats for multimedia components and the browsers used to run them are in a constant state of change.

Similarly, there is ongoing challenge in providing new hardware in learning centers and the workplace so that the latest capabilities of multimedia programs can be taken advantage of. Existing hardware and networks may be perfectly good for certain applications such as word processing or databases, but quite unsuitable for multimedia. Finding the budget to replace current systems in order to run multimedia programs is a problem for managers and administrators and requires that a good business case can be made based on the extra advantages or benefits that multimedia can provide. In some circumstances the cost-benefits are clear-cut (e.g., when money is saved relative to the development of traditional print or audiovisual materials), while in other situations, the payoffs are less tangible (e.g., more interesting materials and hence greater comprehension or retention), and the business case is more difficult to make.

Interactive multimedia define the future direction of computing and telecommunications technology. All of the media described in previous chapters (i.e., electronic publishing, video/television, teleconferencing) will eventually be integrated with current multimedia programs and systems. Thus, the capabilities we see now are quite rudimentary compared to what we can expect in the near future. On the other hand, the capabilities that we have already can create very powerful and effective learning experiences.

Summary of Key Ideas about Interactive Multimedia

- Interactive multimedia bring together all forms of information presentation (text, graphics, audio, video, animation) in a single delivery system, the personal computer.
- Multimedia technology makes it easy to create and capture information in different forms as well as present it.
- Interactivity can range from making menu choices and selecting links to complex input in a simulation or game.
- CD-ROM provides a very cost-effective delivery medium for multimedia programs and a good way to enforce copyright.
- The drawbacks of CD-ROM are that it requires physical distribution, and updates involve a new disc.
- Learning benefits of multimedia include: 1) increased motivation, 2) appeal to a broader range of learning styles, 3) more realistic material, and 4) facilitation of multilingual presentations.
- The four types of multimedia learning environments include individualized settings, group (classroom) settings, EPSS, and public access (kiosk) settings.
- System requirements are an important aspect of multimedia development since there is wide variation in terms of memory, speed, graphic, and audio capabilities across systems.
- Design of a multimedia program requires attention to both the aesthetic and human interface issues.
- Because such a wide range of skills/knowledge is needed to develop multimedia programs, a team approach is recommended.
- To keep the costs and development time of multimedia programs under control, a great deal of selectivity needs to be practiced in terms of what features/capabilities are used.

References

Acovelli, M. and Nowakowski, A. (Nov-Dec 1994) The business practices course: Self-study learning reengineered. *Educational Technology Magazine.*

Blattner, M. and Dannenberg, R. (1992) *Multimedia Interface Design.* Reading, MA: Addison-Wesley.

Hix, D. and Hartson, H.R. (1995) *Developing User Interfaces.* New York: John Wiley & Sons.

Hofstetter, F. (1995) *Multimedia Literacy.* New York: McGraw-Hill.

Howlett, V. (1995) *Visual Interface Design.* New York: John Wiley & Sons.

Kearsley, G. (1994) *Public Access Systems.* Norwood, NJ: Ablex.

Laurel, B. (1990) *The Art of Interface Design.* Reading, MA: Addison Wesley.

Perry, P. (1994) *Multimedia Developers Guide.* Minneapolis: Sams Publishing Co.

Shneiderman, B. (1992) *Designing the User Interface* (2nd Ed). Reading, MA: Addison Wesley.

Silverstein, N. (March/April 1997) Duracell's Integrated Training Approach. *CBT Solutions Magazine.*

Vaughan, T. (1994). *Multimedia: Making It Work* (2nd ed.). New York: Osborne/McGraw-Hill.

10 Simulation, Simulators, and Virtual Reality

S imulators (devices intended to model actual equipment) represent one of the oldest domains of technology-based training. It's also a domain closely associated with military training — simulators of weapons systems (e.g., wooden guns and horses) have been used for many centuries. Of course, over the years, simulators have become increasingly sophisticated, such as the multimillion dollar flight simulators now used to train aircrews. In these programs, the equipment is worth more than the aircraft for which they are training. Although computer-based simulations have only been around since the 1950s when the first computers appeared, the basic idea of simulation as a teaching technique has been practiced for a long time, especially in the form of models, games, or role-playing. While simulators and simulations were initially different entities (one being physical and the other being strictly cognitive), over time the two have largely merged, not in small part due to the fact that both are now computer-based. In this chapter, we will basically treat the two as the same topic, with clarifications as needed.

As simulators/simulations came to be computer-based, their nature and range of application has changed dramatically. While they are still mostly restricted to technical skills training (e.g., hands-on learning to operate equipment), they can now be used to teach any content area including management, sales and marketing, finance, politics, foreign languages, and even training itself. The simulations used for these content areas involve causal models of processes rather than equipment. And, when simulations take the form of games (a relatively small difference), almost any topic or domain is possible as demonstrated by the incredibly diverse range of computer games available. Alas, outside of science classes and medical education, simulations

131

The Role of Simulators in U.S. Air Force Pilot Training

The U.S. Air Force is one of oldest and most prominent users of simulators for training pilots and flight crews — with excellent results. Simulators reduce training time and costs, and produce more thoroughly trained aviators. But there is still room for improvement. A recent study of the Specialized Undergraduate Pilot Training (SUPT) program by the Air Crew Training Division of Armstrong Laboratory identified a number of new directions for simulator use by the Air Force based on emerging technologies and the limitations of the current training.

At present, the Operational Flight Trainer (OFT) which is used in SUPT adequately replicates the basic functions of the training aircraft, but it does not simulate the visual imagery that is essential for effective practice on many flying tasks. Also, students need assistance from instructors to practice in the OFT. The study recommended replacement of the OFT by a Unit Training Device (UTD), a new simulator that would feature a high resolution, wide field of view visual system based on a multiple rear-screen projections system, and/or head-mounted displays capable of producing stereoscopic (3-D) images. The new UTD simulator would also include an automated training guidance system that could adapt training scenarios to the trainee's individual proficiency profile. Furthermore, the UTD would permit simulator networking so that trainees could practice formation maneuvers as well as be able to simulate other aircraft so the student pilot could practice joint maneuvers without the need to interact with an instructor or another student.

The study also recommended other forms of new technology for use in SUPT besides the UPT. This includes a Portable Electronic Trainer (PET) which would be a palm-top PC that would provide each trainee with training support and communications capabilities at any location on a training base. The PET would dock with a desktop training station that could provide computer-based training in a classroom or field setting, including a broad range of instructional simulations that would address various aviation fundamentals (e.g., navigation, landing/take-off procedures, formation flying, etc.). The PET could also be used in actual training flights as an EPSS and to collect data on student performance. The latter would be addressed via an Aeronautical Training Recorder (ATR) which would capture flight data to reconstruct the student pilot's flying actions during debriefing (perhaps using the UDT to replay the flight). Finally, the study recommended a SUBT hub computer system that would integrate all on-line training materials as well as provide a student record database.

Andrews, D. et al. (July/August 1996) Potential modeling and simulation contributions to specialized undergraduate pilot training. *Educational Technology Magazine.*

are used infrequently in classroom or in corporate training rooms. This is because of the difficulty and expense of creating them, the lack of instructor experience with this form of instruction, and, in some cases, the absence of the necessary hardware and software.

Equipment-Based Simulators/Simulation

The classic and most widespread forms of simulators/simulation are devices intended to provide hands-on training for equipment, e.g., weapons, transportation, manufacturing, or medical. Pilot and aircrew training is the most common application for simulation, both in military and civilian sectors. Flight simulators have been in use since the first airplanes were flown; the earliest ones were plywood mock-ups of cockpits with nonfunctional switches, levers, and dials taken from old aircraft. Even in such crude form, these simulators were useful for providing practice in carrying out flight procedures and understanding basic operations. Today, mock-ups are still used in aerospace training (called part-task trainers), although they tend to be exact replicas of the equipment and much of the instrumentation and controls are functional (although controlled by computer models). Most modern aircraft simulators, however, are fully functional and behave in all respects like the actual aircraft — including the computer controls, which are the same as found in the planes. These simulators are mounted on electromechanical support units which are capable of producing the full range of motions experienced in flight and produce appropriate sensations relative to how the plane is flown or weather conditions.

One of the enduring and important questions of the simulator field remains — How realistic does the simulation need to be in order to produce effective training? This is the fidelity issue, and an enormous amount of debate and research has been devoted to it (Hays and Singer). It's a critical question because a high degree of realism costs — the more realistic the simulator, the more expensive it is to develop and build. For the organizations that are buying these simulators, the fidelity issue is one that means a lot of money! If the training can be accomplished with a less expensive, less realistic simulator, that will be more cost-effective. And the research shows that in many cases a high level of realism is not needed — at least not all the time. A large part of a training curriculum can be accomplished with simpler simulations (including those that only involve interacting with a computer system) that still produce effective training. The expensive, high-fidelity simulators can be limited to the final stages of hands-on training that require complete realism.

Driver training is a good example (and there is a variety of driving simulators). For the initial learning of simple driving procedures and rules of the road, a computer simulation will work fine. In order to practice psychomotor coordination and proper reaction times, a simulator that includes the basic controls (steering wheel, gears, brake) is needed, although it does not need to look much like a real vehicle, and the driving environment depicted on the screen can be a relatively simple computer animation. Indeed, the racing car simulators found in game arcades are probably more realistic than needed at this stage. When it comes to the final stages of learning where it is important to develop a physical "feel" for the road and traffic, driving a real car is necessary, but this might represent only a small percentage of the overall training program. Unfortunately, few schools have the simulations/simulators that could be used prior to actual road practice — probably resulting in less effective and more time-consuming driver training.

How could practice in the real equipment be less effective than time spent in a simulation? First, there is the time savings. People can learn specific skills faster in a simulator than the actual equipment because there are fewer irrelevant aspects to draw attention away from the current lesson. Indeed, since a simulation can be programmed to behave in a variety of ways. It is possible to provide much more structured and diverse practice exercises in a shorter period than would appear naturally. For example, think of all the different safety (or flight) incidents that could be simulated that would not likely occur in an actual driving (flying) experience if it was of brief duration. So, the nature of the experience and practice that a person encounters in a simulation can be preprogrammed, making the training more complete as compared to the random nature of actual experiences. Finally, there is the cost factor. Since simulations usually cost a fraction of the actual equipment, it should be possible to give trainees much more simulation practice than if only the actual equipment was available. Even with very expensive flight simulators (which do cost as much or more than the real aircraft), the cost of operating the simulator (electricity vs. fuel) is far less, making it still a cost-effective proposition that allows for more training time.

Instructional Basis of Simulation

The use of simulation and simulators for learning in the workplace has emerged from a practical need to provide hands-on practice. Unlike other instructional

Digital's Remote Access Training Labs

Like most high-technology companies that sell complex equipment, Digital Equipment Corporation faces a continuing challenge to provide hands-on training to its technicians. Providing actual equipment at multiple training locations is not a good option, particularly when the equipment in question is very expensive and/or subject to frequent change. Furthermore, there is always a difficult period when a new product is introduced and few units are available for training a large number of employees who need to know how to install or maintain the new equipment! Conversely, there is also a problem with old equipment that may be difficult to find, yet people still need training on it in order to service customers who have it.

To address these kinds of issues, Digital created a Remote Access Lab (RAL), a single centralized lab with a complete set of hardware and software that can be accessed remotely through an intranet from any location in the world. Virtually any course content requiring hands-on experience can be delivered by the RAL. Applications can be used along with simulations or other exercises built into the system using the actual technology. The RAL provides students with experience in servicing computer products and network environments, while reducing travel expenses and the capital costs of having equipment at multiple sites. The RAL has a full-time administrator who can be on-line with the students and monitor their sessions. The system currently has a capacity to allow 12 to 15 people to access the RAL at one time. Emerging developments such as desktop video and virtual reality promise to expand the capabilities of the RAL to be able to deliver more effective on-line training.

Benson, G. and Cheney, S. (October 1996) Best practices in training delivery, *Technical & Skills Training*.

methodologies, it has not developed historically from theories of learning or teaching. As simulation has come to be used more broadly, however, greater attention has been paid to the underlying theoretical basis (Towne).

Simulation methodology is closely associated with the systems approach since it involves building models, identifying cause and effect (input/output)

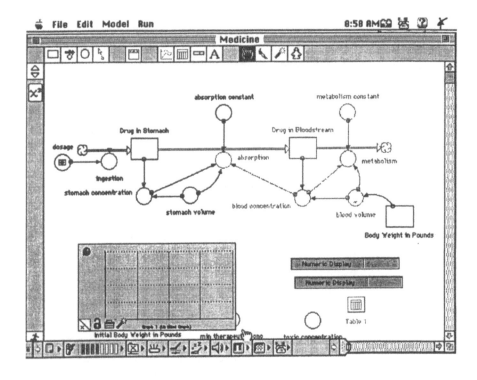

Figure 10.1   The STELLA Program

relationships among variables, defining elements, subsystems, and boundaries, and examining feedback sequences. In fact, many of the early efforts
in simulation derived from systems design efforts, and systems modeling
programs were developed for this purpose. For example, the popular STELLA
simulation program (from High Performance Systems) originated from systems
design work (Figure 10.1). When simulations are developed for instructional
purposes, they tend to reflect many systems design assumptions and concepts.

Simulations are also closely associated with a constructivist view of learning which emphasizes the need for each individual to create his or her own
models of phenomena. So-called "discovery learning" is often associated with
the constructivist view because it is assumed that learners must explore ideas
and hypotheses in order to build cognitive models (including dead ends).
Indeed, this is one of the distinguishing characteristics of simulations — they

provide students the freedom to try things out and make plenty of mistakes doing so. This stands in contrast to instructional theories that attempt to generate a learning sequence that minimizes the number of student errors. One of the best-known advocates of constructivism and technology is Seymour Papert who has focused his attention on the role of programming in children's learning and more recently, on the role of computer games.

Gibbons and his colleagues suggest six major characteristics of instructional simulations:

1. *Coaching* — involves mechanisms which help the student learn from the use of the simulation.
2. *Feedback* — lets the students know how they are doing.
3. *Representation* — the way the simulation conveys information (using different sensory modalities).
4. *Control systems* — determine what and how the student interacts with the simulation.
5. *Scope* — refers to what goals or objectives are achieved by use of the simulation.
6. *Embedded didactics* — instructional messages that the simulation provides about content or process.

The design of an instructional simulation must address each of these different characteristics, whether explicitly or implicitly.

In recent years, a new kind of instructional simulation, based on the framework of goal-based scenarios has emerged, primarily from the work of Roger Schank and the Institute for the Learning Sciences at Northwestern University. The basic idea of goal-based scenarios is one or more simulated tasks with a well-defined goal and success criteria (Keegan). A simulated organization is created and trainees are given a specific set of objectives. For example, Andersen Consulting created "innmasters," a simulation model of a hotel chain to teach consulting skills. In this simulation, the trainee's task was to design, prototype, and document a hotel reservation system for the simulated business. The simulation used graphical, photographic images, audio, and video clips to create a realistic and detail-rich learning environment. Andersen and other companies have used the goal-based scenario framework to develop simulations for a wide range of management topics and skills.

Model-Centered Instruction: Beyond Simulation

Andrew Gibbons, Utah State University, Guest Author

Have you ever wondered if we might have chosen the wrong categories when we divide the world of instruction into classes like tutorial, simulation, job-aid (EPSS), and so forth? As I look beneath the surface of these categories, I am starting to think that we can do better.

We tend to classify instructional products by their mechanism, delivery mode, or experiential features instead of by their instructional essence. "Tutorials" project direct instructional messages and related interactions. "Job-aids" are used to support performance in real environments. "Simulation" encompasses those kinds of products or experiences that resemble reality in some way. These definitions miss the instructional nuance and create classes of instructional tools that are huge, have enormous range and variation, and whose names tell you almost nothing about the instructional characteristics or use of the products.

I suggest the term "model-centered instruction" as a means of describing how simulations, as well as a host of things we don't normally call simulations, are used instructionally in ways that cut across the traditional categories. The central principle of model-centered instruction is that the pursuit of the instructional designer is to: 1) provide real systems or models (of three types: environment, system model, and expert model), 2) with which the learner can interact, 3) during the solution of a problem. Problems are assumed to be selected and ordered by principles that maximize learner benefit.

The traditional notion of simulation is included within the scope of this definition, but this is more: it also encompasses real environments, systems, and expert performances, and relates them (and simulation experiences) to their specific application during instruction. This definition of an instructional type defines things in terms of how they are used in the instructional act. Not a definition of a product type, it defines a specific context and interaction with the learner — a type and structure of experience.

The heresy gets even better. I would like to propose the idea that all instruction is model-centered. Consider the instructor teaching middle school learners about atoms. Since atoms are not visible to the naked eye, the instructor searches for a representation of some sort that is. What are the options? A computerized model? Virtual? 3-D? 2-D? Animated? Noncomputerized? Videotaped? A rapid sequence of photo stills? A handful of isolated stills? And what is the ultimate endpoint of this denaturing of the real into lower-grade representations? A verbal description. If the instructor can find no other means of representing the atom, he or she will resort to verbal explanations of the sort that are common to textbook passages and lectures.

But if you listen to the content of the text or the lecture what do you hear? You hear the description of a system — a verbal model of a system deconstructed by the instructor into verbal representations because that is all that is available. It is not a question of whether we employ model-centered instruction — all instruction is model-centered. It is a question of how far we denature real systems and phenomena into models before we represent them to the learner. The media are simply tools for denaturing real systems and phenomena in order to represent them to the learner.

Model-centered instruction is a useful and important perspective because humans learn and think in terms of (mental) models, not in terms of isolated facts and dissociated elements of knowledge. In the absence of formal instruction, individuals seek experience with real or modeled systems as a source of learning.

What's the impact of this perspective? If it is a correct perspective, it has implications for the way we design instructional experiences, the methods we use to construct them, and the tools we use in doing so. The primary questions of the instructional designer should be, for this learning: 1) what is the appropriate model (or real system) the learner should experience, 2) what is the appropriate level of denaturing for this learner, 3) what sequence of problems should the learner solve with respect to this model or real system, 4) what resources should be available as solving takes place, and 5) what instructional functions to augment the learner's own knowledge and skill should accompany solving?

The method of design changes from analyzing tasks to analyzing environments, systems within environments, and expert behavior toward both. From this analysis comes the specification of a number of problems the designer may select and sequence. From them comes the specification of the learning environments required, the instructional functionalities required, the surface dramatic dynamic required, and the logical structures required (if a computer is being used). The tools of most value to designers are those that can be used to build models of things for learners to interact with — models with a range of denaturings. This means first and foremost that those tools that we have been calling "simulation-building tools" and treating as a luxury become of primary importance, and other representation systems (media) follow in their trail.

Andrew Gibbons is a professor in the Department of Instructional Technology and Director of the Center for the School of the Future at Utah State University in Logan, UT. He has many years of experience in the design of instructional systems along with interests in learning theory. He recently co-authored a book entitled *Computer Based Instruction: Design and Development* (Educational Technology Publications). He can be contacted at: gibbons@cc.usu.edu

Emerging Developments in Simulation: Virtual Reality and Networking

The increasing multimedia capabilities of computer systems allows simulations to provide better visual and auditory presentations. The addition of an expert system to a simulation allows it to offer more sophisticated decision-making or evaluation capabilities. Furthermore, a number of new developments in technology promise to make simulation an even more powerful tool for training: virtual reality and high-speed networking.

Virtual Reality

Virtual reality (Helsel and Roth; Rheingold) refers to the ability to present three-dimensional images and allow sensorimotor interaction with a computer system. The effect of adding these additional input/output capabilities is to give the user the impression of physical involvement or "presence" in a computer environment. To provide these additional sensory capabilities, the user typically has to wear a special helmet or goggles as well as gloves (and possibly a full body suit) which contain displays, sensors, and activators. The additional instruments can also be built into a chair or enclosure that the individual occupies when using the computer. The effect of being totally immersed in the computer program produces a much different kind of interaction than simply using a keyboard/mouse and viewing a two-dimensional screen display. To date, virtual reality has been used in a number of commercially available arcade games and in experimental versions of flight simulators, but it is likely to have a broader impact on computer systems over time.

Networking

Adding networking to simulators also produces a new range of capabilities. The U.S. Department of Defense has explored these capabilities extensively in its Distributed Interactive Simulation (DIS) program, which permits a variety of different weapons systems simulators (both air and ground systems) to be linked in large-scale war games. These exercises involve connecting soldiers and aircrews in many different types of simulators located at different bases in "virtual" battle scenarios where they must coordinate their combat maneuvers electronically in real-time. To do this, DOD developed a common network protocol that allows all simulators to be connected together as well as a dedicated high-speed network capable of transmitting the enormous amounts of data involved in this kind of activity. While this concept has obvious relevance to military training, it could also be valuable in other training settings, such as manufacturing, plant operations, or emergency services which involve similar team and equipment interactions. It could also apply to management or administrative simulations in which teamwork is essential. Of course, simulators/simulations would first need to be used in these training settings, which they are not at present.

Summary of Key Ideas about Simulation/Simulators

- Simulators and simulation are well-established forms of training; however, computers have introduced new possibilities.
- While simulations are most closely associated with military training, they are a general instructional method that can be applied to any skill or content area.
- Simulators have been highly successful in aircrew training, where they can improve performance and provide cost-effective practice.
- A critical issue in simulation is the degree of realism necessary for a simulation to be effective (fidelity).
- Part-task simulators can be cost-effective ways to teach some of the skills required without the expense of a high-fidelity simulator.
- Simulators can provide more effective practice than actual equipment because they can present a wider range of scenarios than would occur in natural settings.
- Simulation is built on a systems approach since it involves building models, identifying relationships, and examining feedback sequences.
- Goal-based scenarios represent a new form of simulation that has been successfully applied to a range of training applications.
- Virtual reality (3-D representations with sensorimotor input/output) offers new opportunities for simulation.
- Networked simulation which allows group/team interaction is another important new development.
- The use of simulation in training has been limited by development costs and equipment availability. Advances in authoring tools as well as declining system costs may change this situation.

References

Andrews, D. et al. (July/August 1996) Potential modeling and simulation contributions to specialized undergraduate pilot training. *Educational Technology Magazine.*

Benson, G. and Cheney, S. (October 1996) Best practices in training delivery, *Technical & Skills Training.*

Gibbons, A., Fairweather, P., Anderson, T., and Merrill, M.D. (1997) Simulation and computer-based instruction: a future view, in C. Dills and A. Romiszowski (Eds.), *Instructional Development Paradigms.* Englewood Cliffs, NJ: Educational Technology Publications.

Hays, R. and Singer, M. (1989) *Simulation Fidelity in Training System Design.* New York: Springer-Verlag.

Helsel, R. and Roth, J. (1991) *Virtual Reality: Theory, Practice and Promise.* Westport, CT: Meckler.

Keegan, M. (1995) *Scenario Educational Software: Design and Development of Discovery Learning.* Englewood Cliffs, NJ: Educational Technology Publications.

Papert, S. (1993) *Children's Machines: Rethinking Schools in the Age of the Computer.* NY: Basic Books.

Rheingold, H. (1991) *Virtual Reality.* New York: Simon & Schuster.

Towne, D. (1995) *Learning and Instruction in Simulation Environments.* Englewood Cliffs, NJ: Educational Technology Publications.

11 Authoring Tools for Learning Technologies

To a large extent, many of the technologies we have discussed in the preceding chapters are enabled through the use of authoring tools — programs that allow developers to create and manage electronic materials. In Chapter 6, we discussed desktop publishing software which constitute authoring tools for creating electronic documents (including graphics and images). In this chapter, we shall examine the tools used for developing interactive multimedia and computer-based training programs, as well as software intended to assist in the design process itself.

Multimedia Authoring Tools

A wide variety of software tools is typically used in creating a multimedia program, although some authoring systems offer many of the capabilities needed. For example, a graphics or photo editor will be needed to create or modify illustrations and images. Many authoring systems include simple graphics/photo editing functions, but they may not be powerful enough to do more advanced operations, in which case a separate editing program will be used. The same applies to video or sound editing.

One of the most commonly used multimedia authoring systems is Macromedia Director (Figure 11.1). This program was originally designed to produce animated movies. While the program has evolved into an all-purpose tool for creating any type of presentation, it is still based on a time sequence of frames and "cast" members to be included. Macromedia Director can be used to import and edit video or sound sequences; the most recent

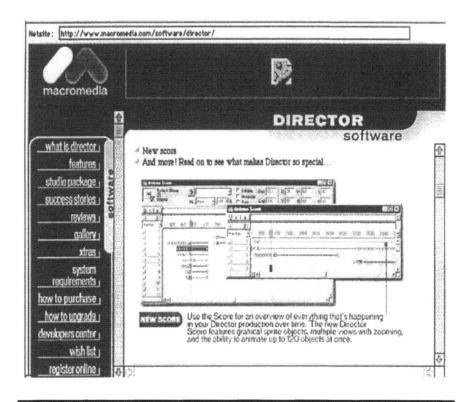

Figure 11.1   Director (Macromind)

versions of the program make it easy to move things to the web via a utility called "Shockwave."

Another widely used multimedia authoring system is Asymetrix Toolbook, which is similar to a desktop publishing program in the sense that programs are developed as a series of pages or screens (Figure 11.2). Multimedia and interactive components (e.g., buttons, links, response messages) can be defined as elements of those screens. Toolbook allows a full range of graphics, images, audio/video, and animation components to be incorporated in a screen, along with control options.

Authoring systems such as Director and Toolbook make it easy to develop a program for CD-ROM distribution. Once the program is complete and fully tested, it can be copied to a CD-ROM using the built-in functions of these tools or the utility software provided with CD-R (CD-ROM recorders) units. For mastering large quantities of CD-ROMs, the multimedia programs would be delivered to a disc manufacturing facility in the form of a master CD-ROM or disk storage media (e.g., cartridges).

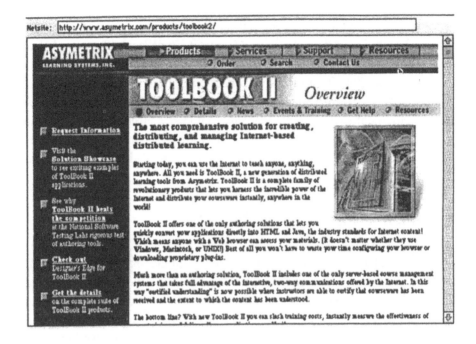

Figure 11.2   Toolbook (Asymetrix)

One important capability of any authoring tool is the availability of a "player" version that can be distributed freely to people to run the programs developed using the authoring tool. In the past, authoring system providers required developers to pay licensing fees for all copies of programs they distributed. Needless to say, this was not a popular concept. Furthermore, now that most programs created via authoring systems are distributed on the web, there is no need for a special run-time version (since web browsers act as universal "players").

Another authoring system concern is determining what kinds of system the run-time program works well with. This includes the type of operating system needed (e.g., Windows version X, Mac, Unix), the kind of multimedia capabilities (color, sound, graphics resolution), and the performance requirements in terms of processing speed and memory size. Ideally, an authoring system can produce programs that can run in a wide variety of different user environments — preferably without the user or developer having to worry about it. Again, this is one of the benefits of distributing programs via the web since browsers take care of these system differences — provided they are properly configured with all the necessary "plug-in" and "helper" utilities.

With the emergence of the web, a whole new generation of authoring systems has emerged for creating web-based multimedia presentations (e.g., Adobe Pagemill, Claris HomePage, Microsoft FrontPage). These tools basically automate the process of writing HTML code and simplify the linking of multimedia files. There is also a collection of tools to help produce JAVA applications (e.g., Aimtech Jamba, Penumbra Mojo) as well as to assist with web site development (e.g., O'Reilly WebSite, NetObjects Fusion). Almost all categories of software are being retooled so they are web compatible. This allows files to be saved in HTML format so they can be directly loaded onto the web.

Computer-Based Authoring Tools

Another category of authoring software that existed prior to the appearance of the multimedia tools mentioned above are Computer-Based Training (CBT) authoring systems. These systems were developed specifically for the creation of instructional programs, including the management of student data and lesson materials. While many of these systems have come to resemble multimedia authoring tools, they still feature basic structures conducive to development of instructional software such as answer judging and branching options.

One of the most widely known CBT authoring tools is the Authorware system marketed by Macromedia (Figure 11.3). Authorware was one of the first graphically oriented systems that allowed authors to define their programs using icons to represent instructional events and components. Icon-Author from AimTech also provides a similar graphics-oriented development environment. On the other hand, authoring systems such as Quest (Allen Communications), CBT Express (AimTech), or Phoenix (Pathlore) are more screen-oriented along the lines of Asymetrix Toolbook (which has a CBT version called Toolbook Instructor). A number of CBT authoring tools have been around for a long time and have evolved from instructional programming languages into the "click and point" systems of the present (Milheim). There are also special-purpose authoring tools available for creating tests or games. An example of the latter is GameMill (Stillwater Media), which specifically allows the creation of instructional games using popular game-show formats such as "Jeopardy" or "Wheel of Fortune."

One of the primary issues in the selection and use of an authoring system (CBT or multimedia) is the extent to which it allows the author to create easy-to-use programs. To some extent, this is up to authors in terms of the user controls and options they provide. On the other hand, the author has

Figure 11.3 Authorware (Macromind)

to work within the confines of what the authoring system allows. For example, the availability of an "undo" option is generally considered to be a very important user function since it allows people to easily recover from mistakes. An authoring system must allow this function to be included in a program and to work properly in the run-time environment. While most current authoring systems do incorporate the most important elements of good user interface design in their functional capabilities, this may not be sufficient for certain applications (such as public access or assistive systems where special user control features need to be addressed).

Another issue is the scope of the instructional strategies which can be implemented using an authoring system. For example, it is often difficult to create a sophisticated simulation or game using an authoring tool, because of the complex displays and user interaction involved. Related to this example is the performance characteristics of the run-time program — does it run fast enough? Simulations normally require very quick animations and response to user actions, but programs developed using authoring systems are often very sluggish in their run-time behavior. For these reasons, a number of specialized authoring tools have been developed for creating simulations, such as the Rapid

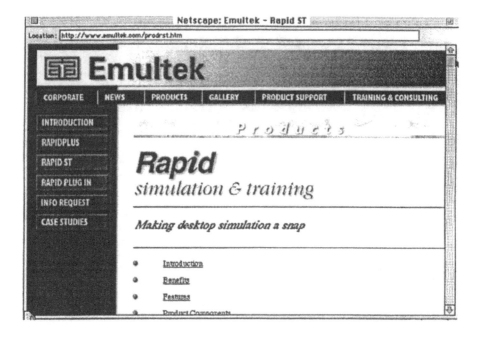

Figure 11.4   Rapid (Emultek)

program from Emultek, which has been used by many organizations for product training and testing (Figure 11.4).

Finally, we should mention the special requirements that EPSS pose for CBT authoring tools. As will be discussed in Chapter 13, some EPSS may be embedded in the applications software they support. This means that the EPSS must be able to run properly in the hardware/software environment of the applications program rather than as a stand-alone program. Similarly, the EPSS usually must not interfere with or monopolize system resources when running — which can be a problem when multimedia elements are involved. Any authoring system used to create an EPSS must work within these performance constraints of the application environment. To the extent that the applications and EPSS are web-based, these issues mostly disappear.

Automated Design Tools

While authoring systems can significantly reduce the development time/costs and improve the overall reliability and quality of technology-based training programs, it is important to keep in mind that most of the development underlying a program (and its success) lies in the design, implementation, and evaluation aspects — which are not affected by authoring systems. Even the best authoring system can probably only impact less than one third of the total workload in a technology project. Not surprisingly, a lot of attention has been given to automating other aspects of the development process — particularly the design tasks (Spector). Many organizations have developed computer-based instructional development systems for internal use that include tools for analysis and evaluation of training, and a number of training development firms market this type of system. For example, Designware from Langevin Learning Services in Canada is an EPSS that supports all aspects of the instructional development process and corresponds to the steps in their model for training system design.

Providing the Foundation for Design Tools

For more than a decade, Omnicom Associates has been developing computer tools for training analysis and design including CEI (the Content Expert Interviewer) and IDeAS (Interactive Design and Authoring System) and most recently, the DesignStation 2000 system. This latter system is an integrated collection of software designed for a laptop computer. It includes capabilities for needs analysis, client presentation and proposal generation, rapid prototyping of print, audio, video, and interactive media projects, collaboration and project-tracking, assessment of financial and educational results, and accessing/adding to a research database. The DesignStation 2000 is based on some fundamental principles of performance technology such as: 1) design is a collaborative and iterative process, not top-down or linear, 2) learning systems are only one possible solution to human performance problems, and 3) designs must be evaluated by their potential for return on investment, both time and money.

Commenting upon the use of automated design tools, Diane Gayeski, one of the partners in Omnicom, says: "In working with organizations to help them adopt instructional design and performance technology workbenches, we have found that the problem lies not in the cost, complexity, or availability of the hardware and software, but in developing standards and philosophies for education and training. Not unlike other software tools, like accounting systems, one needs to have a system in place before computerizing it."

One design task for which software tools are commercially available is storyboarding — the process of identifying the components and sequencing of a multimedia program. Any reasonably large program can involve hundreds of text or graphic components and audio/video or animated sequences, which undergo constant modification as the program develops. Storyboarding tools help keep track of these components and sequences and can print/display a current version of the script at any time. Designer's Edge is an example of such a tool intended specifically for the design of interactive multimedia programs; in addition to storyboarding capability, it also includes functions for needs analysis, writing objectives, and constructing test items (Figure 11.5).

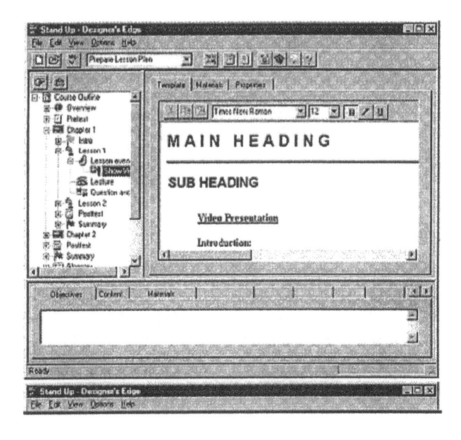

Figure 11.5   Designer's Edge (Allen Communications)

A particularly ambitious effort at automating instructional development is the ID Expert project conducted by M.D. Merrill and Associates at the University of Utah (Zhang, Gibbons, and Merrill). The ID Expert system is an attempt to develop an expert system for instructional development which is based on a new theoretical framework for learning called "instructional transactions." Different classes of instructional interactions are translated into "shells," which can then be used to create knowledge bases for learning. The expert system helps the author fill in all the necessary parameters of the transaction shells (such as objectives, examples, exercises, questions, and answers, etc.) to complete a knowledge base. While the ID Expert system is primarily a research project, it is indicative of the scope and capabilities of future authoring systems. Recently, River Park Corporation has made available two software systems derived from this work, namely, the Electronic Trainer and the Instructional Simulator.

In most large-scale instructional development projects (whether technology-based or not), general application software such as spreadsheet, database, and project management programs are very useful since there are usually many elements to be tracked and accounted for. In fact, most implementation and evaluation aspects of an instructional project can be well handled by an integrated software package that includes these types of functions. However, this requires that instructional development groups have staff who are comfortable using these kinds of programs, as well as support personnel to help install, train, and maintain the software.

Authoring Tools: The Right Stuff

There is a variety of different authoring tools available for the development of technology-based training programs. In fact, there are so many such tools that one of the biggest challenges facing training developers is which software to select. This selection process is made more difficult by vendor claims that their product(s) can do everything and by the fact that it takes considerable time and effort to evaluate a tool to determine if it meets your needs. Indeed, it is usually necessary to actually use a tool in the context of a real application before a good assessment can be made. One shortcut to this evaluation process is to hire a reputable consultant or training development firm that has experience with different authoring tools to provide advice and recommendations. Another possible approach is to ask the vendor of one or more authoring tools to develop a prototype for your training application to demonstrate the functionality and usability of their systems.

Another dilemma associated with authoring systems is the question of who should actually use them. Ideally, they should be used by the subject matter expert who is preparing the instruction. However, this person (or persons) may not have any formal instructional design or training background, and while they can learn to use the authoring tool, they may not be very good candidates to develop effective training programs. Hence, it is more likely that an instructional/training developer will use the authoring tool working with the content and teaching ideas provided by the subject matter expert. But the developer may not have much experience with interface design or interactive multimedia; hence he or she may need the assistance of a software designer or multimedia specialist for these aspects of development. Finally, there are likely to be various system issues to be resolved with the implementation of the program (e.g., compatibility, performance, unexplained errors), which need the attention of a systems analyst or technician. So by the time an authoring system is fully deployed and a program completed, a number of different people are likely to be involved and using the system.

Finally, it is important to recognize that most authoring tools are complex software systems that take considerable time to master. While they are relatively easy to use and can be learned by almost anyone, it usually takes many weeks or months before a person understands them well enough to take full advantage of all their features and develop effective programs. Of course, the learning curve can be speeded up by good training and/or by working with individuals who already have experience with the tool(s). But since most training development projects are on a fast timeline for completion, there is often little time available to develop the mastery of the authoring system needed to use it well. The best solution to this problem, (as well as meeting the need to have a variety of different skills and talents in the authoring team) is to hire experienced consultants or development firms to create technology-based training programs (at least initially). If the programs are successful, the development work can be transferred to in-house staff on a gradual basis as internal personnel become proficient enough, through training and working with outside vendors, to handle it. Most organizations follow this model in their development efforts and use of authoring tools.

Summary of Key Ideas about Authoring Tools

■ While there are many powerful authoring tools, a variety of different programs will be needed to develop technology-based training materials.

■ Authoring systems are based on a number of different models, including movies (e.g., Director), pages (e.g., Toolbook), or icons (e.g., Authorware).

■ An important feature of an authoring tool is that it has a "player" program that can be distributed freely to potential users.

■ Ideally, programs created by authoring tools should run on all major operating systems (e.g., Windows, Macintosh, Unix).

■ CBT authoring systems usually include functions for handling learner input and processing responses.

■ Authoring systems define the nature of the user interface provided in the learning programs, which may or may not be satisfactory for certain applications.

■ Authoring systems also define the scope of the instructional strategies possible, which sometimes may present limitations (e.g., simulations).

■ Development of EPSS presents special considerations for authoring tools because they often need to be embedded in large software systems.

■ Authoring tools speed up the development process but do not affect design and analysis work.

■ Automated instructional design systems attempt to improve the effectiveness of the analysis and design aspects.

■ Other application software (such as spreadsheets, database managers, and project tracking) can improve the efficiency of instructional development projects.

■ Selecting the right authoring tool(s) is difficult; obtaining the help of an experienced consultant/vendor in this area is recommended.

■ Authoring involves various kinds of skills/background and is best done by a team.

■ Because there is a significant learning curve associated with any authoring tool, relying on consultant/vendor support initially is recommended.

References

Gayeski, D. (May-June, 1995) DesignStation 2000: Imagining future realities in learning systems design. *Educational Technology*.

Milheim, W. (1994) *Authoring Systems Software for Computer Based Training*. Englewood Cliffs, NJ: Educational Technology Publications.

Spector, J.M. (1993) *Automating Instructional Design: Concepts and Issues*. Englewood Cliffs, NJ: Educational Technology Publications.

Zhang, J., Gibbons, A., and Merrill, M.D. (1997) Automating the design of adaptive and self-improving instruction, in C. Dills and A. Romiszowski (Eds.), *Instructional Development Paradigms*. Englewood Cliffs, NJ: Educational Technology Publications.

12 Knowledge — the Critical Resource for Global Success

The way in which organizations use technology to manage knowledge has quickly emerged as the single most important discriminator between success and failure in this intensely competitive global economy. Nonaka and Takeuchi, authors of *The Knowledge-Creating Company*, confidently predict that a company's ability to create, store and disseminate knowledge will become absolutely crucial for staying ahead of the competition in quality, speed, innovation, and price. Only by developing and implementing systems and mechanisms to assemble, package, promote, and distribute the fruits of its thinking will a company be able to transform knowledge into corporate power.

Thomas Stewart, in his classic *Intellectual Capital: The New Wealth of Organizations*, writes, "Simply put, knowledge has become more important for organizations than financial resources, market position, technology, or any other company asset." Knowledge is now seen as the main resource needed for performing work in an organization since all the organization's traditions, culture, technology, operations, systems, and procedures are all based on knowledge.

Intelligent organizations recognize that only one asset grows more valuable as it is used — the knowledge skills of people. Unlike machinery that gradually wears out, materials that become depleted, patents and copyrights that grow obsolete, and trademarks that lose their ability to comfort, the knowledge and insights that come from the learning of employees actually increase in value when used and practiced.

155

Yet, in most companies the use of technology to manage its knowledge is totally uncharted territory. Managing know-how may be different from managing cash or buildings, but "managing the intellectual investments of a company needs to be treated every bit as painstakingly" (Stewart).

Organization-wide knowledge is needed for several reasons: a) to increase the abilities of employees to improve products and services, thereby providing quality service to clients and consumers; b) to update products and services; c) to change systems and structures; and d) to communicate solutions to problems. The ability to manage knowledge (and learning how to manage that knowledge) should become the primary job of every worker.

Organizations, therefore, need to learn how to manage the "mechanics of knowledge;" just as in the industrial era they learned how to manage the "mechanics of production." In this chapter we will discuss the steps and strategies for managing knowledge, now only possible through the adroit use of technology. To better understand and apply these steps, however, we first need to look at two ways of clarifying what knowledge is: a) hierarchy of knowledge, and b) levels of knowledge.

Hierarchy of Knowledge

Unless and until a company determines what "knowledge" is and what type of knowledge is important, there is no way it can manage its corporate knowledge. Obviously, not all knowledge is of equal value to an organization and needs to be classified according to its importance. Figure 12.1 presents such a hierarchy or continuum of knowledge. As one goes up the hierarchy, there is an increase in breadth, depth, meaning, conceptualization, and value. Let's briefly examine each level.

Data include texts, facts, interpreted images, and numeric codes that have not yet been interpreted, have no context, and therefore do not yet have meaning.

Information is data that is imbued with context and meaning, whose form and content are useful for a particular task after having been formalized, classified, processed and formatted.

Knowledge, in turn, is a body of information, principles, and experience to actively guide task execution and management, decision-making, and problem solving. Knowledge is that which enables people to assign a meaning to data and thereby generate information. Knowledge enables people to act and to deal intelligently with all the information sources available.

Figure 12.1   Hierarchy of Knowledge *(Source:* **Liebowitz, J. and Beckman, T. [1998]** *Knowledge Organizations: What Every Manager Should Know.* **Boca Raton: CRC Press.)**

Expertise is the application of knowledge in an appropriate and effective way in order to achieve results and performance.

Capability represents the organizational capacity and expertise to produce a product, service, or process at a high level of performance. Capability requires the integration, coordination, and cooperation of many individual and team efforts. It is more than just current performance, but also the ability to learn, to innovate, and create.

Types of Knowledge

Knowledge can also be classified as types of knowledge:

- Knowing what information is needed (*"know what"*)
- Knowing how information must be processed (*"know how"*)
- Knowing why the information is needed (*"know why"*)
- Knowing where information can be found to achieve a specific result (*"know where"*)
- Knowing when the information is needed (*"know when"*)

Wiig further distinguishes knowledge on four conceptual levels:

1. Goal setting or idealistic knowledge — vision, "care-why" knowledge or self-motivated creativity
2. Systematic or "know-why" knowledge to acquire systems understanding
3. Pragmatic of "know-how" knowledge to acquire advanced skills
4. Automatic (tacit) or "know-what" knowledge for routine working.

In developing the knowledge-based system or corporate memory of the organization, it important to recognize the different distinctions and values of these types of knowledge and where they fit in the hierarchy and needs of the organization.

Systems Model for Managing Knowledge

Let us now look at a comprehensive systems approach for the management of corporate knowledge through technological support. A cohesive knowledge-management system involves five stages as knowledge transitions from source to use: 1) knowledge acquisition, 2) knowledge storage and mining, 3) knowledge analysis, 4) knowledge sharing and dissemination, and 5) knowledge application and validation (See Figure 12.2).

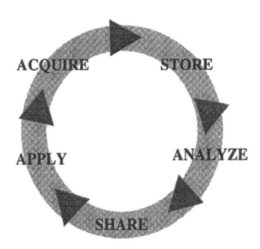

Figure 12.2   Knowledge Management System

For organizations to manage their knowledge effectively and efficiently, each of these five components must be ongoing and interactive. The management of knowledge should be continually subjected to perceptual filters as well as to both proactive and reactive activities. As we noted in Chapter 3, the management of knowledge is at the heart of building a learning organization. Successful learning organizations systematically and technologically guide knowledge through each and all of these five stages. Let's now carefully examine the knowledge management system, exploring the role of technology in each of the stages.

Knowledge Acquisition via Sourcing and Creating

An increasingly overwhelming volume of knowledge from a variety of sources all over the world is required for people to adequately perform their work. Only through the careful application of sophisticated technology can needed knowledge be gathered. In determining the appropriate technology for knowledge acquisition, it is important to consider how the data/information will be later retrieved by different groups of workers for performing their job tasks. Functional and effective knowledge storage systems (as we will note in the following section) should be categorized around the following elements:

- Learning needs
- Work objective
- User expertise
- Function/use of information
- Location — where and how is the information stored

Organizations acquire information and build their knowledge base by a) sourcing information from both external and internal sources, and b) creating new knowledge.

Sourcing Knowledge

Internal Collection of Knowledge

One of the major complaints of workers is that so much of their knowledge is never tapped by the organization. Often, companies are startled to learn how much intellectual capital is present in the brains of their own employees, what Nonaka calls tacit knowledge.

Tacit sources of knowledge include individual employee's expertise, memories, beliefs, and assumptions, all of which can be of high value to the organization. These tacit sources are usually difficult to communicate or explain, but will result in tremendous benefit to companies. McKinsey & Company (see Chapter 19) has demonstrated remarkable capability and creativity in collecting the knowledge of its workers.

External Collection of Knowledge

The pace of change is so rapid today that no single organization can ever gain control of or dominate all effective operating practices and good ideas. To be a marketplace leader, an organization must look outward for constant improvement and new ideas. The old school of thought, which held that "if it isn't invented here, it can't be any good" is a curse in today's high-velocity markets. Organizations don't need to invent what others have learned to do well. Today's rallying cry for companies is "acquire, adapt, and advance!" Companies can "rope in" information externally through a large number of methods:

1. Benchmarking from other organizations
2. Attending conferences
3. Hiring consultants
4. Reading print materials such as newspapers, e-mail, and journals
5. Viewing television, video, and film
6. Monitoring economic, social, and technological trends
7. Collecting data from customers, competitors, and resources
8. Hiring new staff
9. Collaborating with other organizations, building alliances, and forming joint ventures.

Creating Knowledge

Knowledge can also be created through a number of different processes "ranging from ingenious innovation to painstaking and elaborate research." It can also come from the "uncanny ability people have to see new connections and combine previously known knowledge elements through complex inductive reasoning" (Wiig) The knowledge created through problem-solving, experimentation, and demonstration projects can often be the most valuable form of knowledge for organizations.

Knowledge Storage

Knowledge storage is not new. In fact, the concept first emerged in the 1980s. Once data were in place and catalogued, managers could help themselves to whatever slice of the company's data pie they needed at that moment. The idea sounded good. In practice, however, the size ("data deluge") and complexity of the resulting data warehouse meant that costs involved were very high — too high for everyone except a few banks and airlines.

Within the last few years the concept of data warehousing has reappeared, and is now quickly traveling around the world. Why? The answer has to do with competition as well as the dramatic reductions in cost and increase in power of today's computers. Thus comprehensive knowledge repositories, the "on-line, computer-based storehouses of expertise, knowledge, experience, and documentation in which knowledge is collected, summarized, and integrated across all information sources are emerging in organizations around the world. The expensive proprietary mainframes of first-generation data warehouses are being replaced by much cheaper UNIX systems. Web technologies such as Java, JavaScript, CGI, and Active X, as well as GB Oracle databases, have also emerged. (See the case study of AMS in Chapter 19 for a description of some of these technologies)

A knowledge storage system enables an organization to contain and retain knowledge, so that it becomes company property and doesn't go home at night or leave the organization when the employee leaves. Unfortunately, knowledge (also referred to as "intellectual capital"), though far more important than physical material, is usually scattered, hard to find, and liable to disappear without a trace because it is not stored. Storage is obviously important, but what knowledge should be stored?

What Knowledge to Store

Stewart proposes five general categories under knowledge should be stored:

1. *Corporate Yellow Pages* — capabilities of employees, consultants, and advisors of the organization; e.g., who speaks Thai, who knows JavaScript, who has worked with client A?
2. *Lessons Learned* — checklists of what went right or wrong that might be applied to other projects; leverage what has been learned in the past rather than losing the recipe

3. *Competitor and Supplier Intelligence* — continuously updated company profiles and news from commercial and public sources and wire services; call reports from sales people; attendees' notes from conferences and conventions; an in-house directory of experts, news about regulations
4. *Company Experiences and Policies* — process maps and work flows, plans, procedures, principles and guidelines, standards, policies, performance measures, shareholder and customer profiles, products and services (including features, functionality, pricing, sales, and repair)
5. *Company Products and Processes* — technologies, inventions, data, publications and processes; strategies and cultures, structures and systems, as well as organizational routines and procedures that work

How Knowledge Should Be Stored

Knowledge is only clumsy data unless it is coded and stored in a way that makes sense to individuals and the organization. Too many organizations remain overwhelmed and inundated with vast amounts of data that clutter up the information highway. An organization cannot learn from the information if it is irretrievable, distorted, fragmented, or inaccurate.

To determine what data can be used, the organization must decide what is of value and then code the data based on learning needs as well as on organizational operations.

The knowledge stored should be easily accessible across functional boundaries. It should be structured and organized so the users can find concise information quickly. Store the knowledge not only by topical categories, but also based on learning needs of staff, organizational goals for continuous improvement, and user expertise. Finally, the knowledge stored should be updated so it remains accurate and valid.

In addition, the knowledge acquired should be stored in a way that it would be easy:

- For workers to decide which co-workers could have the knowledge needed for a particular activity
- For workers to decide which co-workers would be interested in a lesson learned
- For a worker to submit a lesson learned to the corporate memory (there should be a well-defined criteria for deciding if something is a lesson learned, how it should be formulated, and where it should be stored)

Capturing and Storing Knowledge at Cigna

Cigna Corp., a leading insurance company, knows that excellence comes from making more knowledgeable choices. Cigna recognizes that there is significant latent knowledge and expertise in the organization, but, until recently, did not have a good means of extracting and publishing this know-how. Cigna gave home-office managers the job of building and maintaining a knowledge base — basically a collection of checklists, rules of thumb, formal guidelines for risk assessment, and names of experts. It was installed into the same software that every underwriter used to process applications. If a nursing home in California wants insurance, the custom-built software tells where the nearest geological fault line is and how dangerous the company's experts consider it to be. When new information comes in, the manager/knowledge editor evaluates it and, if he thinks it's good, changes the database. Every underwriter instantly incorporates this new best practice.

Challenges in Storing Knowledge

It is important to remember that knowledge storage involves technical (records, databases, etc.) and human processes (collective and individual memory, consensus). As organizations become physically and geographically more spread out as well as more specialized and decentralized, the organization's storage system and memory can become fragmented and corporate benefits of the knowledge can be lost. And as work becomes more computer orientated, the informating needs of different organizational specialists become potentially available across functional boundaries. Networked information technology must be utilized so that fragmented information can be reinterpreted and readily exchanged internally and externally.

Given the fact that new technology is increasingly able to store and provide more information to the organization's members, consideration must be given to the potential of data deluge or information overload. The amount of information stored should not exceed the organization's capacity to adequately process the data.

Knowledge Analysis and Data Mining

Over the past 30 years, organizations have become skilled at capturing and storing large amounts of operational data. Unfortunately, until recently, we have not seen corresponding advances in techniques to analyze this data — to reconstruct, validate, and inventory this critical resource. Manual analyses with report and query tools remain the norm, but this approach fails as the volume and dimensionality of the data increase. New approaches and tools are therefore needed to analyze very large databases and interpret their contents.

The latest development in analytical tools is data mining. Data mining enables organizations to find meaning in their data. By discovering new patterns or fitting models to the data, employees can store and later extract information to better develop strategies and answer complex business questions. Software is being developed that can analyze huge volumes of data and identify hidden patterns within that data. Whereas OLAP (on-line analytical process) can answer the questions managers ask, data mining software answers the questions managers didn't even think of!

There are several data mining tasks (classification, regression, clustering, summarization, dependency modeling, and change and deviation detection) as well as data mining methods such as:

- Decision trees and rules
- Nonlinear regression and classification methods
- Example-based methods
- Probabilistic graphical dependency models
- Relational learning methods and use of intelligent agents

Data Mining Tools

A number of data mining tools are being developed for navigating data, for discovering patterns and creating new strategies, and for identifying underlying statistical and quantitative methods of visualization. Platforms to support these tools, techniques to prepare the data, and the ways to quantify the results are also emerging. For example, visualization products include AVS/Express, SGI MineSet, and Visible Decisions Discovery. To integrate data mining products, DataMind and IBM's Intelligent Miner are extremely helpful.

AMS Analyzes Data for U.K. Bank

AMS and a banking client in the U.K. to sought to analyze more than 267,000 records of 132,000 customers to find correlations among customer behavior, demographics, and profitability. To look at these large volumes of data across 152 business dimensions, AMS used sophisticated statistical and visualization techniques. These include scatter plot matrices and parallel coordinate plots in conjunction with brushing, cutting, and high-dimension rotation.

Applying these techniques provided analysts with new perspectives on their data. They rotated and twisted a multidimensional image to search for relationships among groups of data to discover new patterns or trends. In this case, AMS was able to find correlations between channel usage (e.g., checks, branch visits, ATM site), age, and occupation to help the bank quickly zero in on the most profitable segments of bank customers. High payoff uses also include credit card scoring, risk analysis, product profitability analysis, retail site planning, and manufacturing.

Data mining is being utilized by a growing array of organizations, including:

Retailers — The universal adoption of EPOS and the spread of loyalty cards is fueling rapid need for knowledge analysis. Key benefits are the ability to understand customers' buying behavior and to rapidly identify unprofitable lines. W.H. Smith, for example, is reported to have weeded out 20,000 of its least profitable products as a result of knowledge analysis, and the biggest U.S. retailer, Wal-Mart, is investing in the world's biggest data warehouse to handle data on customer buying patterns from its 2900 stores.

Financial services organizations have long seen the potential of knowledge analysis to obtain an integrated view of their customers. High payoff areas include highly targeting database marketing-and-risk analysis. Capital One Financial Corp has virtually revolutionized the credit card business by using data mining to do highly sophisticated customer profiling and targeting. As a result, it has been able to develop a portfolio of literally hundreds of target credit card products.

Manufacturers — Some of the latest data mining techniques are being used by manufacturers. Downtime is expensive — such as when a paper roll breaks in a paper mill — but by analyzing the data from previous stoppages, the combination of circumstances in which downtime occurs can be predicted with accuracy. This knowledge is then matched in real time against the current operational conditions so that avoidance action to be taken — a prevention rather than a cure.

Telecom companies generate huge volumes of customer and operations data. These call data are increasingly being analyzed in ever more sophisticated ways by telecoms operators to determine competitive pricing, develop new pricing tariffs, and to design highly segmented marketing campaigns. Improving network utilization is another application of knowledge analysis technologies, e.g., analyzing how many calls do not get through to specific customers because they don't have enough lines to satisfy the demand.

Knowledge Sharing and Dissemination

The sharing and transfer of knowledge is the sharing and transfer of power — a corporate capability indispensable for the corporate success. Simply stated, knowledge needs to be disseminated accurately and quickly throughout the organization to where it is needed, or the company fails.

The retrieval of knowledge may be either controlled (by the memory or records of individuals or groups) or automatic (when situations trigger memories, ways of doing things, etc.). Weick warns that as a result of the transformational nature of the storage and retrieval process, the normal integration of human memory, the impact of perceptual filters, and the loss of supporting rationales, information that is retrieved from organizational memory may bear little resemblance to what was originally stored. It is therefore very important for an organization to develop a corporate memory and design processes in a manner to ensure accurate and timely knowledge retrieval.

Just-in-time access of required information leads to extension of an individual's long-term memory and reduces the working load memory. The corporate knowledge base consolidates information into a central location, thus liberating an individual's working memory from such menial data as resource location. This creates the conditions for the rapid sharing of knowledge and sustained, collective knowledge growth. Lead times between learning and knowledge application are shortened systematically. Human capital will also become more productive through structured, easily accessible and intelligent work processes.

Ways of Sharing Knowledge

Knowledge sharing involves both the organizational and technological movement of information and knowledge, and can be distributed within an organization both intentionally and unintentionally.

Organizational Ways of Transferring Knowledge

Organizational, interpersonal ways of transferring knowledge include:

1. Individual written communication (memos, reports, letters, open access bulletin boards)
2. Training (internal consultants, formal courses, on-the-job training)
3. Internal conferences
4. Briefings
5. Internal publications (video, print, audio)
6. Tours (especially for large, multidivisional organizations with multiple sites that are tailored for different audiences and needs)
7. Job rotation/transfer
8. Mentoring

Technological Modes of Transferring Knowledge

A comprehensive, wide-scale transfer of knowledge, however, can only proceed through the intelligent use of technology, so that the knowledge can be available anywhere, anytime, and in any form. Information communications software, including electronic mail, bulletin boards and conferencing, allows for interactions between members, both person-to-person and among dispersed groups. It also provides an electronic learning environment where all members have equal access to data and are able to communicate freely.

If all the organization's personal computers are networked through the mainframe with relevant external systems, any person can take part in gathering and transferring knowledge. Remote access to national and global knowledge networks can be made available within the organization at any time. Sharing information on a virtual real-time basis and encouraging wider access to information involves:

- Creating on-line databases that can be used across functional boundaries
- Hooking into on-line databases and electronic bulletin boards external to the organization such as universities and other learning centers
- Installing an electronic mail culture where its use is widespread
- Using electronic data interchange to create comprehensive electronic network systems

Newer search engines now allow for searches of all files located on a LAN or WAN using search criteria found in many web browsers. Current groupware offers the incorporation of expert system or decision support systems into a standard graphical user interface and gives individuals access to knowledge that has been imported into the expert system from the external environment (as well as capturing knowledge that exists within the organization, both congenital and experiential).

Barriers to Sharing and Transferring Knowledge

According to a recent survey conducted by The Knowledge Management Network, three major bottlenecks were identified that prevent companies from sharing and transferring knowledge:

1. Critical business processes are available only to a few people.
2. Knowledge is not available at the place and/or at the point-in-time when needed.
3. Transfers and restructuring increases the difficulties in securing knowledge.

Several companies, however, have made great strides in developing a knowledge sharing system. Let's look at three of them.

Andersen Consulting— Knowledge Xchange is used, which allows over 17,000 professionals located in 47 countries to utilize the system daily to access knowledge bases and share knowledge.

Chevron uses intranets, groupware, data warehouses, networks, bulletin boards, and videoconferencing as tools for distributing stored knowledge. Knowledge bases, lists of experts, information maps, corporate yellow pages, custom desktop applications, and other systems are also utilized.

GTE — uses CYLINA (Cyberspaced Leveraged Intelligent Agent) that acquires knowledge through interactions with large numbers of users. This is supplemented by Auto-FAQ, which is a question–answering system that helps users retrieve knowledge from CYLINA.

Knowledge Application and Validation

Stewart notes that systematic application of "intellectual capital creates growth in shareholder value." This is accomplished through the continuous recycling and creative utilization of the organization's rich knowledge and experience. Technology is needed to gain optimum application value from corporate knowledge. The ability of a company to provide customer service through diagnosis and troubleshooting is a good example of knowledge application and validation as shown by the following example.

Hewlett-Packard created an electronic network to manage and distribute knowledge to keep up with customers' demands for speedy service on a global scale. H-P's customer response network supports 199 technical support staff, mostly engineers, whose job is to keep customers' computer systems up and running. These systems are the customers' central nervous systems. If they go down, they must be fixed fast. When a client reports a problem, the electronic messages go automatically to one of four hubs around the world, depending on the time of day. Operators get a description of the problem and its urgency, type the information into a database, and zap the findings into one of 27 centers, where it might be picked up by a team specializing in that area. The database is shared by all the centers and is "live" — that is, whenever, an employee works on a file, it is instantly updated, so every employee has identical information about each job at all times. If the first center can't solve a problem quickly, it follows the sun — California, Australia. HP managers are involved in moving the work around the net; it is seamless.

A well-developed knowledge system allows a company to put its best people on the front line while still keeping their expertise available to the entire organization. Charles Paulk, chief information officer of Andersen Consulting, notes that, because of Knowledge Xchange, "When one of our consultants shows up, the client gets the best of the firm, not just the best of that consultant." Among the benefits of its knowledge management system, Paulk lists the following:

1. Savings (Andersen saves millions in FedEx bills alone)
2. Ability to more easily tap into colleagues' knowledge
3. Helps Andersen work globally
4. Maps corporate brainpower
5. Allows coping with growth and staff turnover (the faster newcomers can learn what the institution knows, the faster they can contribute to its success)

Technological Support for Knowledge Management

Now that we have examined the five-stage model of knowledge management, let's look at how to provide an overall technological support system. Beckman suggests four sequential but overlapping stages in assuring that the knowledge is collected, stored, and shared:

Stage 1 — Establish an Installed Networked IT Infrastructure for all Employees

There should be a networked IT platform installed across the organization to support the knowledge systems. All employees should be equipped with a workstation that supports complex computational, informational, and communication needs. Every worker should be able to communicate electronically with all other employees, both as individuals and collaboratively in groups. Powerful system navigation and information exploration tools that use flexible key word search, hypermedia, dynamic visual querying, and tree maps should be available. Employees should also be provided a range of standard office automation software, including text processing, presentation graphics, spreadsheets, rational DBMS, Web browsers, and e-mail.

Stage 2 — Create Enterprise-Wide Data, Object, and Knowledge Repositories

At this stage, organization-wide relational and object models as well as data dictionaries should be created and regulated. Existing on-line data need to be reformatted before being inserted into the company-wide databases. Smart data-entry templates should be available to check for validity, consistency, and reality. The knowledge repositories should include software to translate media into text.

Stage 3 — Automate and Enable Operations, Management, and Support Activities

With the assistance of electronic systems, all operations should be automated within the organization. For example, in marketing and sales, the technology would be able to better match products to customers' needs as well as increase profit margins through improved pricing.

Stage 4 — Develop Integrated Performance Support Systems and Knowledge Discovery and Data Mining Applications

In stage 4, centers of expertise should be formed that would be responsible for collecting, storing, analyzing, and distributing knowledge. The centers would train and certify workers in their specialties and provide qualified workers and consulting services both on-line and in-person.

The most dramatic and successful applications will come as a result of the deployment of Integrated Support Systems that provide workers with coordinated task information, advice, training, job aids, references, and administrative resources. These centers of expertise have several roles:

1. Create, research, improve, and manage the knowledge repository
2. Set and enforce standards, methods, and practices
3. Align and coordinate interests with related centers
4. Assess work force competency and performance
5. Identify gaps and remedy deficiencies in the content and processes of the knowledge repository
6. Provide training and consultancy services
7. Supply competent workers to staff projects and processes

Accelerating Performance Through Learning and Knowledge at Ernst & Young

Tom Solomon, Senior Manager
Professional & Organization Development Group
of Ernst & Young

As Ernst & Young fully enters the knowledge economy, it recognizes how technology is radically redefining competitive dynamics. The only way we can prosper is through the collective and individual learning power of the people who make up our company. Our learning has never been more vital than it is now. One of the key ways Ernst & Young is striving to ensure its competitive advantage is through a program called LEAP (Learning Environment to Accelerate Performance). The following are some of the underlying principles on which LEAP is based:

1. *Performance* — Ernst & Young believes that as a learning organization it must first and foremost be focused on performance at three levels — individual, team, and organization. Serving as a human performance advisor to a variety of clients requires that learning be linked strategically with the goals of the firm.

2. *Transformation of Learning* — Learning has traditionally meant the process of helping individuals acquire skills. Ernst & Young has developed a new definition of learning — one that moves beyond employee development into a broader and more vital context: helping the firm to respond to competitive issues, helping the firm to reinvent itself. In this context, corporate learning includes helping groups build and share a vision, helping teams collaborate, helping individuals and groups make better decisions, and empowering people to do their jobs better by providing the learning products they need at the point of performance. This new definition challenges us as learning advisors to come up with new processes and solutions for the constituencies we serve.

3. *Technology Solutions* — Technology has enabled Ernst & Young to offer new and exciting tools to accelerate the performance of our people. Satellite-based distance learning programs allow Ernst & Young to transmit knowledge around emerging issues in key industries like health care, telecommunications, and financial services. Web-based training allows users to develop basic skills in

tax or change management at their own pace and at the time most convenient for them. We are currently developing web-based communities of practice that allow people to share information and opinions with geographically disparate team members. Even our classroom training is affected by technology through the use of laptop PCs, video conferencing, and software tools. This reinvents the classroom to better represent the technology-enabled Ernst & Young workplace.

4. *Knowledge Reuse* — Ernst & Young is a leader in knowledge management through its wide use of Lotus Notes and its corporate intranet. The firm has also invested in creating and leveraging massive amounts of knowledge in the form of best practices, example deliverables, and thought leadership. Having this collateral (much of it in digital form) provides an incredible source of content knowledge. LEAP enables us to re-use this content by incorporating it into solutions that are specifically created for learning. Learning solutions help provide a better context for knowledge by tying it to the firm's sense-making and decision-making processes.

5. *Learning Environment* — To build the firm's overall learning environment, Ernst & Young has developed advisors and advisory services, new design and development processes, and new technologies that reflect its vision of learning. It has developed these both internally and via alliances with innovative vendors. Examples of these include a comprehensive learning management system co-developed with a software vendor, a distance-learning network hosted by a distance-learning company, and collaborative learning environments developed by a knowledge management software company.

Ernst & Young sees many exciting challenges ahead for the learning industry. Global companies will continue to expand, technology will continue to change, and companies will need to reinvent themselves in the workplace. These factors will create the need for larger and more complex learning solutions. Only companies that have redefined learning in its proper broader context, linked it to their business strategy, and created an effective learning infrastructure will be better able to continually innovate to serve their markets. Ernst & Young plans to be one of those companies.

HRIS/HRMS

An important category of software relevant to knowledge management in the workplace is Human Resource Information/Management Systems. These systems (usually developed and implemented by vendors) provide automated solutions to a broad range of human resource functions including: recruiting and job postings, benefits enrollment and processing, career planning and personnel development, performance appraisal and tracking, and aspects of payroll.

Historically, HR information and transactions have been paper-based, involving enormous numbers of memos, policy/procedure handbooks, and forms. In the past decade, fax and telephone answering systems have sped up the processing of HR information, but it has basically remained a paper-based system. However, once the information came to be stored in computer databases and most employees obtained access to computers, it was a natural evolution for HR transactions to become increasingly computer-based. With the advent of LANs and intranets, this evolution toward on-line dissemination and processing of HR information has progressed significantly to the point where almost all HR functions in organizations are done electronically.

This evolution toward HRIS/HRMS has great advantages for both employees and management. Employees can easily access up-to-date benefits information and, in many cases, the software can personalize the information for the specific employee based on their personnel record. On-line forms for claims or requests can include helps and error-checking routines that reduce mistakes and improperly completed submissions (saving time for all). Information about job postings, new business opportunities, changes to benefits, etc., can be distributed quickly and uniformly to all employees via e-mail or bulletin postings. Routine HR transactions (e.g., processing of benefits claims or payroll) can be performed very quickly and at lower cost. More complicated tasks, such as performance appraisal, recruiting, or salary planning, can be done much more efficiently because the relevant data are available in databases for examination and manipulation via spreadsheets and specialized analysis tools.

Most large organizations are in the process of moving their HRIS/HRMS systems to their LANs and intranets to make information dissemination and data collection easily accessible to all employees. In an industry white paper prepared by NetDynamics Inc. (http://www.ihrim.org/resources/whitepapers), they describe how a number of their clients are doing this, using the PeopleSoft HRMS and their web server software. Two companies at the forefront in HRIS/HRMS technology for knowledge management are AT&T and Boeing.

PeopleSoft Personnel Technology at AT&T

With a heterogeneous computing environment, AT&T's strategy is to deliver functionality from numerous disparate systems through a common web-based user interface, while consolidating existing custom applications such as recruitment into PeopleSoft. AT&T kicked off its intranet efforts with four self-service applications:
■A recruitment application for posting job requisitions, updating resumes, and viewing internal job postings
■A personal data maintenance application in which employees can maintain information such as address, e-mail ID, and military status
■A manager reporting application for retrieving employee demographic data such as salary band groupings
■An address update application for updating mailstops and processing employee work address changes.

The web-based self-service applications eliminate the need to fax hard copy forms and process written change requests, reducing costs and generating cleaner data for the organization. AT&T employees are now able to review and update their personal information at their desks with virtually no training or software installation. Managers can better oversee the recruiting cycle and research salary information, and the company has in place an enhanced, integrated system for managing costly HR information flow processes.

HRIS/HRMS at Boeing

Boeing Corporation developed a "My Information" web site for employees on the company's intranet. The application starts from a central Boeing People Services page. After logging on, the employee enters a personalized general services page, categorized into personal information, career information, and financial information. The employee may view or change records including: W-4 form, skill inventory, home address, work phone and mail stop, emergency contact, education, monitored training, and paycheck information. Some updates to information occur directly or, in some cases, are indicated as "pending" until the appropriate batch system processes the request. Boeing is extending the My Information web site to support more concurrent users and has ambitious plans for added features, including job applications for internal transfers, additional salary and contact information, payroll deductions, stock option balances, automated check deposit, and directory information updates for their existing electronic mail system.

As these two examples illustrate, one of the main outcomes of implementing HRIS/HRMS in an organization is to empower employees in terms of their access to personnel and benefits information. It also provides managers with much more detailed tracking data about employee and organizational performance. While all of this is generally positive, it does raise important considerations about privacy and security of sensitive personnel data. As we discussed in Chapter 4, advances in technology bring with them ethical issues that must be addressed along with the performance improvement outcomes.

Knowledge Must Move to Be of Value

The greatest challenge for the corporate leader is to create an organization that can redistribute its knowledge. Knowledge is useless unless it moves. By finding ways to make knowledge move, an organization can create a value network — not just a value chain. The ability to intelligently manage knowledge through technology is absolutely essential for an organization's success.

In the remaining chapters of Part III, we will look in more depth at how EPSS (Chapter 13) and networks — including Internet, intranet, the world wide web, and LANs (Chapter 14) — can be used to manage knowledge. In Chapter 15, our guest author, Douglas Weidner, will present how knowledge engineering uses these technologies to provide just-in-time knowledge to its users.

References

Beckman, T. (1998) Expert systems applications: designing innovative business systems through reengineering, in *Handbook on Expert Systems*. Boca Raton: CRC Press.

Liebowitz, J. and Beckman, T. (1998) *Knowledge Organizations: What Every Manager Should Know.* Boca Raton: CRC Press.

Liebowitz, J. and Wilcox, L. (Eds.) (1997) *Knowledge Management and Its Integrative Elements.* Boca Raton, FL: CRC Press.

Marquardt, M. (1996) *Building the Learning Organization.* New York: McGraw-Hill.

Meiklejohn, I. and Duncan, S. (Autumn, 1997) Mining the data mountain. *Fast Track.*

Nonaka, I. and Takeuchi, H. (1995) *The Knowledge-Creating Company.* New York: Oxford University Press

Stewart. T. (1997) *Intellectual Capital.* New York: Doubleday.

Sveiby, K. (1997) *The New Organizational Wealth.* San Francisco: Berrett-Koehler.

Tapscot, (1996) *Digital Economy.* New York: McGraw-Hill.

Weik, K. (1993) The nontraditional quality of organizational learning. *Organizational Science.*

Wiig, K. (1997) Role of knowledge-based systems in support of knowledge management, in *Knowledge Management and Its Integrative Elements.* Liebowitz, J. and Wilcox, L. (Eds.). Boca Raton, FL: CRC Press.

13 Electronic Performance Support Systems

One of the most significant developments in technology, learning, and knowledge management during the past decade has been the emergence of the Electronic Performance Support System (EPSS). An EPSS is any computer-based program that assists people to do their jobs more effectively. This ranges from simple on-line help functions and tutorials that explain how something works to sophisticated multimedia demonstrations involving video. It can also involve the use of conferencing or network capabilities to provide human assistance either in asynchronous or real-time modes. In fact, just about all forms of technology that we have discussed in the previous chapters could be a component of an EPSS.

EPSSs have come to the forefront of HRD practice for a number of reasons. One is the increased emphasis in organizations on performance technology and "just-in-time" training. Indeed, the whole idea underlying an EPSS is not to provide training as such, but the specific information that a person needs to do a task/job exactly when they need it. For example, if a person is learning how to use a new accounting system, the EPSS provides as much detail about each step of the system as is needed as he or she goes along. Different individuals will need varying degrees of information depending on their background. Similarly, the EPSS should help people who already know how to use the system but have forgotten how a particular function works (or may never have had to perform a certain function before). An EPSS is an on-line version of job aids — a performance technology that has long been in use for technical training.

Another factor that caused EPSS to emerge as a knowledge-management technology is the fact that many employees now have access to a computer in their workplace. Since the computer is already there and used by the individual as a routine part of his or her job, it makes sense to provide support using the same computer. Information provided via traditional means (i.e., printed documentation or meetings) is less convenient and not always available like an EPSS. Furthermore, many EPSSs are developed as part of the computer systems used to perform the job so they are more closely tied to job functions than information provided in off-line form, which is often too general and unconnected to be really useful.

An EPSS Is a Many-Splendored Thing

While all EPSSs have a similar purpose (i.e., improving job performance), they take many different forms depending on the work setting. Furthermore, a successful EPSS will evolve over time as the nature of the job, the skills of employees, and the technology change.

The prototypical form of an EPSS is an on-line help system (Duffy, Mehlenacher, and Palmer). On-line help systems can be added to existing software applications used by employees in their jobs to make these programs easier to use and reduce errors/problems. In many cases, help functions are needed to compensate for poor design of the original software — particularly in terms of human factors or job design considerations. In newer programs (including most commercial software applications), help functions are a built-in feature. Most help functions provide detailed explanations of how program options work, the meaning of screen displays, or the exact nature of input expected. Figures 13.1, 13.2, and 13.3 show the main help function for the Apple Macintosh Operating System, which provides information via topics, indexing, and search. More advanced help systems can show demonstrations of procedures or take the user through a series of steps to be performed for a task. The most sophisticated help systems (called "wizards," "advisors," or "assistants") can actually recommend a course of action and automatically carry it out upon user request. For example, Figure 13.4 shows the wizard screen from Microsoft Powerpoint that automatically creates slide formats, and Figure 13.5 shows an Assistant from ClarisWorks that will set up an address database for the user.

While most EPSSs are an integral part of another software system, they can stand alone — as is often the case in equipment-related or manufacturing environments. For example, a workstation containing an EPSS on CD-ROM might be placed on a factory floor to help assembly workers and technicians operate or repair complex manufacturing machinery. An EPSS running on a laptop computer may be provided to engineers to aid them in installation

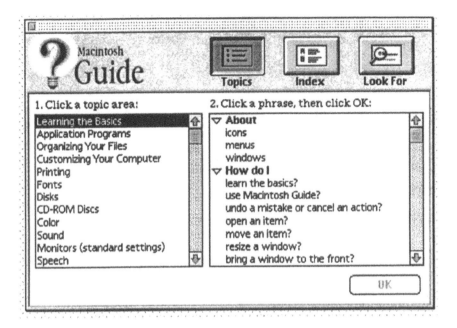

Figure 13.1   Mac Guide Screen with Topic Option Selected

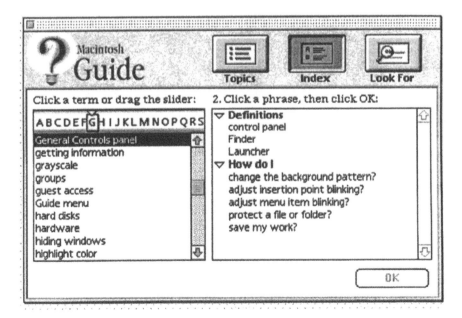

Figure 13.2   Mac Guide Screen with Index Option Selected

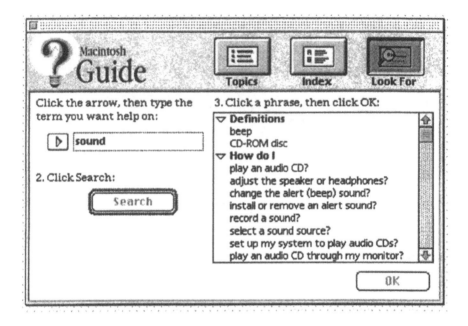

Figure 13.3   Mac Guide Screen with Search Option Selected

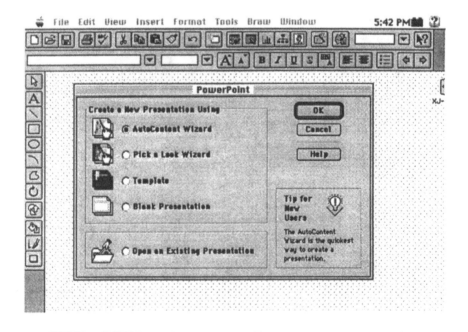

Figure 13.4   Wizard Screen in Microsoft Powerpoint

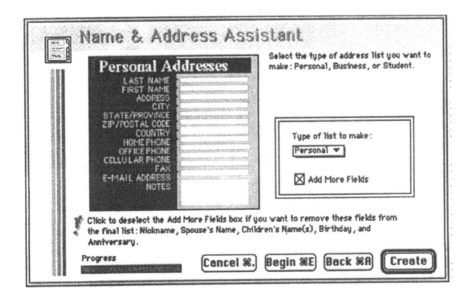

Figure 13.5 Assistant Screen in ClarisWorks

or service activities in the field. The U.S. Department of Defense has developed a broad range of self-contained EPSS units to assist soldiers in the operation and maintenance of weapons systems and combat support equipment. In most cases, such stand-alone EPSSs feature multimedia capabilities (e.g., audio/video sequences, graphics, animation), since they can take advantage of computing resources dedicated to the EPSS. EPSS which is an integral aspect of other software systems is generally more constrained in its capabilities to avoid overloading the primary system.

Another common type of EPSS is an information database located on a network (either a LAN or Internet/intranet). In many organizations, employees need details about products, procedures, regulations, market data, production records, etc., to do their jobs. The information needed can be internal or external (proprietary or public). While one might consider that access to this kind of information would be a routine consideration in a modern organization, the fact is that it is usually not until an EPSS is created. Furthermore, putting on-line information in a format that is easy for a particular person to access and use in his or her job is a specialized undertaking, and the heart of designing an EPSS.

The most intriguing type of EPSS, as far as organization learning is concerned, is the expert system (Durkin; Liebowitz). An expert system contains the collective wisdom of experts on a specific topic or process, e.g., how to troubleshoot faults in a system or how to make the best financial decisions. Since many job tasks involve difficult judgment or decision-making skills that take years to acquire, this kind of expertise is very valuable. Furthermore, it tends to represent key knowledge assets of an organization which may disappear when veteran employees leave. By building an expert system, such expertise can be shared across the organization and be available to new employees who lack the experience represented by the expertise contained in the system. While expert systems are probably one of the most powerful forms of an EPSS, they are also the most difficult to construct and hence least often developed.

The EPSS Design Process

While the computer technology used to deliver an EPSS tends to get most of the attention, the design process associated with the EPSS is what determines its effectiveness. Underlying any successful EPSS is a very careful analysis of job functions and user needs — basic performance technology activities. In the initial design stages of creating an EPSS, a lot of interviews and observation are needed to identify the specific information that employees need to do their jobs. This analysis also includes an examination of why people fail to do their jobs well and where the major performance problems exist. Such an analysis often uncovers fundamental job design issues as well as management and organizational problems. In fact the design work for an EPSS may suggest that the major obstacle to improved job performance is poor management practices or organizational procedure/policies.

Another important aspect of EPSS design is a thorough inventory of employee skills and background knowledge. Since an EPSS is intended to be a personal tool to help each employee do his or her job better, it must accommodate the different needs of individuals. Employees may have difficulty doing their jobs because they lack the skills/knowledge expected of them — which an EPSS may or may not be able to compensate for. If this gap is too large, the issue becomes one of employee selection or placement, i.e., finding more qualified individuals to do the job (which may or may not be an option). On the other hand, every person is going to be variable in skills and knowledge; the purpose of an EPSS is to address these deficiencies for each person if possible. In designing an EPSS, the most common or typical deficiencies are taken into account first.

Boeing Corporation Uses Expert System to Prepare CAD/CAM Training

Boeing Corporation has applied expert systems to a number of areas including the training development process itself. In the area of CAD/CAM, Boeing decided to use an expert system to reduce the long learning curve associated with CAD/CAM software by teaching the thinking processes and strategies of expert users. To do so, they used a methodology known as Knowledge Analysis and Design Support (KADS) to identify, capture, and analyze the thinking processes of CAD/CAM experts. KADS is an alternative to task analysis for work situations that are primarily cognitive in nature (such as software use). The knowledge models constructed as part of KADS were created by skilled analysts through structured interviews with experts over a 3 to 4 month period. The KADS analysts then worked with course developers to integrate the CAD/CAM thinking practices into the training curriculum. This included: specific examples that illustrate certain tasks, tests to check whether students can perform tasks at the desired proficiency level, particular points in the curricula where expert thinking processes should be used, and a set of knowledge models to be used in the course. The end result of the KADS work was to develop curriculum materials that were more complete and effective. In one field test comparing 70 students trained in CAD/CAM according to expert practices with 30 employees using traditional curricula, the majority of those in the KADS-based training were able to complete a design task successfully, while less than half the traditionally trained group were able to do so.

G. Benson and S. Cheney, (October 1996) Best practices in training delivery, *Technical & Skills Training*.

Once the details of job performance and employee skill/knowledge levels are well understood, the design issue becomes what kind of EPSS would be appropriate: on-line help, information database, expert system, or something else? The problem at this stage is to decide the best format for assistance. Also taken into account at this juncture is the way the EPSS will be delivered, i.e., as an integral part of an existing system, a separate program running in conjunction with another system, or completely stand-alone, perhaps with its own hardware/software. The choice of delivery system affects the EPSS capabilities (e.g., multimedia features) and how easy it will be to access on the job. Considerable attention must also be given to the design of the user interface (discussed in Chapter 9) to ensure that the EPSS is easy to use.

Actual development of the EPSS involves creating the information to be provided by the EPSS (such as help messages, database entries, expert advice). This can be a challenging task since the information needed may come from a variety of difference sources within or outside the organization. In many cases, much of the information needed may never have been made explicit before (this will certainly be true for expert systems), and so extracting and verifying information is time consuming and difficult. Furthermore, determining how to present the information so it is immediately useful to EPSS users takes much work and many revisions. Indeed the entire development process of an EPSS requires a lot of iterations as users try things out and the design is fine-tuned (Stevens and Stevens).

Implementing an EPSS: the Long and Winding Road

As difficult as designing an EPSS is, it is the implementation process that is the most tricky. For one thing, developing an EPSS almost always requires the collaboration of different departments or groups within an organization — groups that may never have worked together before. This typically involves the information systems/systems engineering and the training/HRD/technical publications groups. This collaboration requires very close contact and sharing of detailed methodology over a relatively long period of time. The systems group must explain how the system works and the training/publications group must explain Performance Technology and the intent of the EPSS. There must be a high level of trust between the two groups and excellent communication. To ensure this kind of interaction, it is common to create an EPSS task force consisting of team members from the different groups involved (including of course, the client).

Bellcore:
Developing EPSS for Telephone Companies

Bellcore (a spin-off from Bell Labs) has been involved with the development of EPSS for various telephone company systems, many of which were initially created years ago and run on older mainframe computers. For example, Bellcore developed an EPSS in the form of an on-line help system for the MARCH program used to configure electronic switching equipment. After conducting a task analysis of switching operations, they were able to identify certain kinds of information needed to do the job that were not easily available — and hence formed the basis for the help system. Emphasis was placed on providing multiple pathways to help topics as well as extensive indexing and search capabilities. Interestingly, the MARCH help system included video segments demonstrating procedures, but the video was replaced by still photos which were found to be as effective and improved system performance.

In a more ambitious project, Bellcore developed the AdVisor help system, which provided context-sensitive information relative to the screen or field the user is in. When the user activates the help system, it provides information specific to the current screen or field, including explanation of what is expected and what the response options are. The AdVisor system was especially valuable for new users because it eliminated the need for lengthy training beforehand. A third help system developed by Bellcore was MediaVantage. Unlike the other two systems, this one did not need to run on a mainframe system and could take advantage of the multimedia capabilities of the Windows environment. Each screen of the application program was linked to a help page that contained an image of the screen, with pop-up boxes to explain the purpose of every field or button. This help system was found to be very easy to use and reduced the need for application training and documentation.

Across all of the EPSS efforts, the Bellcore team reports some common lessons: 1) meet with your customers and the system developers early and often to be sure you're on the right track, 2) be sure to know who is doing what in the development of the EPSS — it requires very close coordination with the systems developers, and 3) be sure to match the right kind of EPSS with the user need and the system capabilities.

From G. Larson, (May/June 1997) Enhancing performance through customized on-line learning support. *Technical & Skills Training.*

One of the reasons the implementation process is complicated and collaboration is so critical is because the target system is typically under constant change. The EPSS will be effective only if it corresponds to the system as it currently works. When a change is made by the systems group, it must be immediately reflected in the EPSS. If the EPSS was simply another system function, it could be handled by the systems group itself. Each addition to the EPSS requires job and user analysis, i.e., the application of performance technology. It is possible to assign a performance technologist to the systems group for ongoing maintenance of an EPSS, but this requires long-term commitment on the part of management and the understanding that an EPSS is not a one-shot effort. Alas, this level of commitment and understanding is often lacking.

Another aspect of the collaboration process is the initial and ongoing involvement of the client for whom the EPSS is developed. Conducting the task/user analysis needed to design the EPSS takes an enormous amount of time and cooperation from the user population. Managers, and employees themselves, may be reluctant to provide the time needed unless they fully support the EPSS effort and understand the design process. Furthermore, continued involvement of employees will be needed over time as the EPSS is revised to be more effective and reflect system changes. Under ideal circumstances, users are given some (or all) of the responsibility for developing and maintaining an EPSS they use — making it in their own best interests to have an effective system. In essence this is the same as involving all employees in a TQM process — of which an EPSS could be an important component because it represents improvements to work productivity.

Beyond the collaboration considerations are the complex decisions associated with the delivery system for the EPSS. If the EPSS is an integral part of the application system, it must not adversely affect performance of the system, yet still exhibit good performance. This becomes a major issue if the EPSS involves any kind of multimedia components (especially video or even graphics) because these components tend to require a lot of storage space and processing capability. On the other hand, if the EPSS is to take a stand-alone form, performance is not likely to be a problem, but the costs of providing additional equipment may be. To the extent that an EPSS is to be available to many employees at different locations, a large number of dedicated PCs could be needed. However, given the relatively low cost of computers compared to the value of the increased productivity that an EPSS can produce, justifying the expense of additional machines needed should not be too difficult if a cost/benefits study is conducted.

Redefining the Nature of Work

While the overall context for development of an EPSS is to improve job performance (and hence organizational productivity), the use of such systems tends to actually redefine the nature of work activities. Once system designers (and employees themselves) realize that an EPSS can allow people to accomplish a lot more, they start to develop/request more sophisticated on-line capabilities. Furthermore, job responsibilities and roles change because a given individual is likely to be able to carry out a more diverse range of tasks, with more competency, than without the EPSS. For example, a clerk who traditionally handled simple accounting functions, may be able to accomplish a broad array of financial reporting tasks with the help of a good EPSS. Similarly, a technician who previously performed simple repair operations may now be capable of complex diagnostic and maintenance activities through the use of a troubleshooting EPSS. Even managers and professionals can change the nature or level of their work functions with an EPSS (particularly in the form of expert systems).

Table 13.1 lists a set of attributes of EPSS developed by Gloria Gery, one of the leading EPSS experts. These attributes illustrate that an EPSS can address many aspects of job behavior ranging from simple things (e.g., show evidence of work progression) to the very complex (e.g., reflect natural work situations). The list also indicates why such EPSS take such diverse forms — no single EPSS is likely to accomplish more than a few of the functions listed. While the ideal EPSS would address all of these attributes, in practice it is necessary to identify which aspects of job performance need the most attention and develop an EPSS to support those aspects. Clearly, development of an EPSS requires the Performance Technologist and HRD staff to make decisions about the relative importance of different employee tasks and their significance to organizational mission and effectiveness. This is also why the creation of an EPSS needs top management participation — it involves prioritization of organizational goals.

It is obvious that we have a great deal of room to grow with EPSS technology. If even a few of the attributes listed in Table 13.1 were implemented in an EPSS for most jobs in an organization, there would be a tremendous improvement in workplace performance. Of course, associated with such an effort would be a careful analysis of job tasks and performance problems, which itself would be of fundamental benefit. For this reason, development of Electronic Performance Support Systems is a critical mechanism for re-engineering an organization toward increased learning, productivity, and effectiveness.

Table 13.1   Attributes of Electronic Performance Support Systems

1. Establish and maintain a work context
2. Aid goal establishment
3. Structure work process and progression through tasks and logic
4. Institutionalize business strategy and best approach
5. Contain embedded knowledge in the interface, support resources, and system logic
6. Use metaphors, language, and direct manipulation of variables to capitalize on prior learning and physical reality
7. Reflect natural work situations
8. Provide alternative views of the application interface and resources
9. Observe and advise
10. Show evidence of work progression
11. Provide contextual feedback
12. Provide support resources without breaking the task context
13. Provide layers to accommodate performer diversity
14. Provide access to underlying logic
15. Automate tasks
16. Provide alternative knowledge search and navigation mechanisms
17. Allow customization
18. Provide obvious options, next steps, and resources
19. Employ consistent use of visual conventions, language, visual positioning, navigation, and other system behavior

EPSS and the Organizational Learning Process

An EPSS can also be a very valuable tool in building a learning organization because it provides the necessary info-structure needed to help organizations learn more effectively. Here are some reasons EPSS enhances a company's chances of coming a learning organization:

1. *Performance-Centered Design.* EPSS is designed to enable an individual to reach the required level of performance in the fastest possible time and with the least personnel support. The system's design includes such things as embedded knowledge, the ability to structure the flow of work, and adaptability to individual performers.

2. *Performance.* The EPSS leverages the worker's inherent intellectual and social skills by presenting information, knowledge, advice, and support at the moment of need.

Expert Systems for Retirement Planning: T. Rowe Price

An area where EPSS has many applications is financial analysis and planning. Since there are so many variables involved in financial decisions, the benefits of an expert system are great. T. Rowe Price, headquartered in Baltimore, is one of America's largest investment management firms. An important aspect of their business offerings is retirement planning, for which a number of expert-system based software tools (e.g., a retirement planning analyzer, IRA analyzer) are made available for customers. These tools ask users to provide background on their particular financial situation and goals by answering questions and choosing among options. The program then presents to them the best options based on the rules of the expert systems. Such systems allow T. Rowe Price customers to explore various options on their own time and without the pressure of real-time interaction (allowing for a completely private session).

Expert systems for consumer decision-making and evaluation of options is an important, but relatively untapped application. The T. Rowe Price software is an interesting prototype that we may see much more frequently as organizations look for ways to distribute their expertise throughout their customer base. For more information, see <http://www.troweprice.com>

3. *Individual Learning.* As the worker uses the EPSS, he or she can learn in three ways: 1) the worker may change his or her behavior after receiving negative or corrective feedback from the system; 2) the worker may review EPSS modules on the job just before using them; and 3) the worker may review EPSS modules off the job when mistakes would not be dangerous and costly.
4. *Generation of New Knowledge.* The worker will develop new techniques, methods, and procedures on the job that were not part of the original knowledge base. In this way the person creates new knowledge.
5. *Knowledge Capture.* As individuals or teams gain new knowledge, the EPSS captures it through some formal process (through mail messages, shared databases, interview, with expert workers, etc.).

Summary of Key Ideas about EPSS

- EPSS can take many forms, ranging from simple textual helps to complex multimedia demonstrations or on-line conferencing.
- An EPSS is not designed to be instructional in nature but to provide specific job information in the work context.
- EPSS can be an integral component of a software system or a entirely separate program that runs independently.
- Expert systems are a special kind of EPSS that provide assistance based on expertise captured in a program.
- The most important aspect of EPSS design is a thorough analysis of user behavior and job/task requirements.
- Development of an EPSS usually requires collaboration across different organizational functions.
- Initial and continued involvement of the user community is essential in the development and implementation of an EPSS.
- Decisions about how to implement an EPSS must take into account system performance and availability considerations.
- To be successful, an EPSS must be constantly updated and maintained.
- Implementation of an EPSS often redefines employee roles and responsibilities and, in turn, organizational structure.
- EPSS designers should know what changes in job performance they wish to achieve.
- EPSS provides a valuable info-structure for building a learning organization.

References

Benson, G. and Cheney, S. (October 1996) Best practices in training delivery, *Technical & Skills Training*.

Duffy, T., Mehlenacher, B., and Palmer, J. (1992) *On-line Help: Design and Evaluation*. Norwood, NJ: Ablex.

Durkin, J. (1994) *Expert Systems: Design and Development*. NY: Macmillan.

Gery, G. (1991) *Electronic Performance Support Systems*. Tolland, MA: Gery Performance Press.

Larson, G. (May/June 1997) Enhancing performance through customized on-line learning support. *Technical & Skills Training*.

Liebowitz, J. (1990) *Expert Systems for Business and Management*, Englewood Cliffs, NJ: Yourdon Press.

Stevens, G. and Stevens, E. (1995) *Designing Electronic Performance Systems*. Englewood Cliffs, NJ: Educational Technology Publications.

14 Networks: The Internet, Web, and LANs/WANs

While computer networks have been around since the 1960s, it is only within the past decade that they have become such powerful aspects of organizations. Today most large corporations and agencies could not function without computer networks. Similarly, learning via computer networks has become a vital element of post-secondary education and is just beginning to strongly influence the training world. The Internet and the web can take most of the credit for this development, although LANs (local area networks) and WANs (wide area networks) are very important as well.

The Internet: a Network of Networks

Let us begin by understanding what the Internet really is: a network of networks. The Internet is basically a telecommunications protocol, TCP/IP, that allows networks to share data files and specifies how files/messages can be transmitted from one network to another. Every network that the Internet connects can run under different operating systems and computer hardware but can share data by virtue of using the same TCP/IP format. While the count is constantly changing, the Internet connects approximately 500,000 networks around the world of varying sizes and complexity. The networks of some organizations and institutions are quite gigantic, consisting of many LANs and thousands of computers (e.g., consider the U.S. Department of Defense or a multinational corporation). On the other hand, a small business could have its own network, consisting of just a few personal computers!

Each network that is part of the Internet has a unique electronic address that takes the form: name.class (such as: ibm.com, irs.gov, gwu.edu). The class component of the address indicates a particular type of organization, such as a commercial business (.com), a government agency (.gov), or a post-secondary institution (.edu). Addresses outside the U.S. usually have a country code following the class type, e.g., ibm.com.jp for IBM Japan. The address may also include a server identification such as: system1.ibm.com. Each server represents a separate network at that organization. Users on a given network have a unique account name that is pre-fixed to the address by the "@" sign as in: user1@system1.ibm.com. This relatively simple addressing system allows millions of people to interact and transfer information across the thousands of computer systems connected via the Internet.

While the programs that run on all the computer systems that are interconnected via the Internet are totally different, there are some common data transfer functions that each site supports. The most important of these is electronic mail (e-mail). Each mail message must have certain components so it can be transmitted over the Internet to its intended recipient(s). Furthermore, most e-mail programs provide common functions such as carbon copy (CC), automatic reply, file attachment, and receive notification. Even though people are using different e-mail programs on different systems, the messages can be sent back and forth via the Internet because of these common formats specified in Internet protocols.

Another important Internet function (at least historically) is news groups and listservs. Both of these functions provide a means for broadcasting messages to people. Once a person has "subscribed" to a news group or listserv, he or she automatically receives all the messages that anyone posts to the news group or listserv address (e.g., ourgroup@gwu.edu). Effectively they work like electronic newsletters in which anyone can provide information. There are thousands of different news groups and listservs, many devoted to very specialized topics, and most open to anyone who wants to subscribe to them. They represent one of the many ways people use the Internet informally for learning — by joining groups on topics of interest and sharing ideas with people.

On-line conversation in the form of real-time interaction (usually called "chatting") is another function supported by the Internet (Internet Relay Chat or IRC). The significance of real-time interaction as a form of computer conferencing was discussed in Chapter 8. Of all the communication functions provided by the Internet, chat is probably the least significant because we have a much better way to accomplish this form of interaction — the telephone! On the other hand, once this interaction is upgraded to include video images (i.e., desktop videoconferencing), real-time communication via the Internet becomes very popular and more useful. Once desktop video becomes feasible for more people, we can expect to see this Internet function used heavily.

Finally, the Internet makes it possible to transfer files (called File Transfer Protocol or FTP) and access data files on different systems. Such files can be text documents, graphics, numerical data, programs, or multimedia (audio/video) in nature. To the extent that people need to exchange or view information across systems, file transfer and access is a critical function of the Internet. However, these functions have often been a bit obscure on the Internet, requiring users to deal with the complexities of different operating systems and file formats. Software utilities such as "Fetch" and "Gopher" made things much easier, and the appearance of the web simplified the process of file access and transfer immensely.

The World Wide Web: a New Era in Computing

The world wide web, which just emerged in the mid-'90s, ushered in an entirely new era in computing. The web is "just" another Internet application, but one so powerful that it has transformed the way the Internet (and computers overall) are being used. It has made accessing information (data files) very easy and provides a superb method for the delivery of interactive multimedia.

At its core, the web is simply a file format called HTML (Hypertext Markup Language) and a class of programs called browsers that can read them. HTML specifies things like new paragraphs, font size, and centering. In most respects, it is just a text formatting language — except for two important differences. It allows the specification of links to other files (the hypertext capability), and it allows file components to be multimedia in nature (i.e., graphics, audio/video, animation). The links specified in an HTML file can be to any machine and file directory on the Internet using its Internet address (e.g., www.gwu.edu/myfiles/index.html). The multimedia components must be in one of several standard formats for each media that are recognized by all browser programs so they can be run properly.

Browsers (such as Netscape or Explorer) exist for every type of machine, and all can execute basic HTML files. To handle the various multimedia components, additional programs called "plug-ins" are provided that allow a given browser to play a certain format of audio/video or display a specific kind of graphics. Today, most browser software includes plug-ins for all major multimedia formats when initially purchased and installed. However, as vendors develop additional kinds of formats, new plug-ins are developed and can be added to existing browsers. This approach allows the web to have the flexibility of rapid growth and change, without requiring everyone to change their browser software daily! In most cases, new plug-ins can be downloaded automatically from their developers' web site, making the addition of new functionality to a browser relatively easy. For example, there is a lot of development going with different formats for compressing audio and video (so-called "streaming" audio/video), resulting in new plug-ins for these new formats.

One additional capability of the web that is not part of the HTML language itself (at least current versions) is input processing. In order to process input (and hence support interaction), it is necessary to write routines in a programming language such as PERL or Visual Basic which reside on the server along with the HTML files and are executed by the browser when input is required. The most common type of input processing is called "Forms" and is used for submission of text responses. Any kind of interactivity found in other programs (e.g., games, simulations) can be programmed into a web document. New web-specific programming languages such as JAVA and other authoring tools (see Chapter 11), have been developed and are being used for more advanced interaction.

While HTML is currently the principal format used for web documents, there will likely be many others in the future. For example, VRML (Virtual Reality Markup Language) exists for the development and running of three-dimensional animations and simulations (discussed in Chapter 10). Furthermore, all of the vendors who make browser software have developed HTML extensions that only their browser supports in an attempt to give their products a competitive edge. Conceivably, organizations could develop their own proprietary web languages for use only on their own intranets. Of course, this would diminish the power of the web for common access to documents across all networks and systems.

Use of the Web for Learning

In the short time the web has been available, it has had tremendous impact on education, especially post-secondary institutions (Berge and Collins; Khan). Since most colleges and universities have plenty of computing facilities and already make extensive use of the Internet, the web was readily accessible. Almost every large (and even small) post-secondary institution in the U.S and elsewhere has a web site containing pages created by faculty, staff,

Table 14.1   Sampling of Educational Web Sites

www.ncook.k12.il.us — Glenview District 34 Schools
www.imsa.edu — Illinois Math & Science Academy
gseweb.harvard.edu — Harvard School of Education
www.chem.lsu.edu — Louisiana State University Chemistry Department
www.ed.gov — U.S. Department of Education
grad.usda.gov — U.S. Department Agriculture Graduate School
www.au.af.mil — U.S. Air Force Academy
ed.info.apple.com — Apple Classrooms of Tomorrow (ACOT)
www.iat.unc.edu — University of North Carolina Institute for Academic Technology
hub.terc.edu — Technology Education Resources Center (Science/Math)
www.ael.org — Appalachia Educational Laboratory (AEL)
ncrve.berkeley.edu — National Center for Research in Vocational Education
www.upenn.edu/gse/cpre — Consortium for Policy Research in Education
www.ecs.org — Education Commission of the States
school.discovery.com — Discovery/Learning Channel
www.webcom.com/~vschool — The Virtual School
www.gsn.org — Global Schoolnet Foundation
www.educom.edu — EDUCOM
www.aspensys.com/eric2/welcome.html — ERIC
www.rit.edu/~easi — Project EASI
www.enc.org — Eisenhower Clearinghouse for Math & Science
novel.nifl.gov — National Institute for Literacy
www.artsednet.getty.edu — Getty Center for Education in the Arts
www.nea.org — National Educational Association
www.naschools.org — New American Schools
www.ets.org — Educational Testing Service

and students. In addition, many public (K–12) schools have their own web sites for the use of students, staff, teachers, and parents. Plus local, state and national agencies involved in education, as well as vendors and non-profit organizations with interests in the education domain, also have web sites. A selection of such sites is listed in Table 14.1.

In schools, web sites typically contain some or all of the following:

- General information about the school (mission, location, maps of campus/buildings, staff profiles/listings, enrollment, financial aid, clubs, athletics, housing, etc.)
- Course-related information such as schedules, syllabi, outlines, study guides, readings, and ancillary materials (including graphics or audio/video)

- Administrative information for students or faculty including assignments, registration or grading procedures, promotion/hiring policies, meeting schedules/locations
- Student course work (e.g., papers, reports) and faculty writing (conference papers, articles, reviews)
- Access to on-line libraries and on-line help (academic, computer, counseling)
- Miscellaneous information such as bus schedules, daily cafeteria menus, school newsletters, bookstore specials, etc.

In some of the more advanced schools, almost all of the resources and information needed by students, staff, and faculty are available on-line from their web site — which effectively turns them into a virtual institution. To the extent that e-mail and computer conferencing (or some other type of teleconferencing) is used for interaction among students, faculty, and staff, the need for traditional classrooms and face-to-face meetings is greatly reduced. Indeed it is quite possible to have all information and interaction through the Internet/web, creating a full-fledged distance-learning environment. Very few educational institutions are interested in completely dispensing with regular classroom contact, however, so web sites are presently used to support and supplement traditional instructional regimes. Plus, only a small percentage of the faculty is likely to be comfortable doing all their teaching on-line, since this requires quite different methods and techniques than classroom instruction (Harasim et al.).

For students, the web represents an exciting and powerful resource (Campbell and Campbell). It provides a way to find information for course assignments and projects without the need to physically visit a library or buy books. It also allows them to post their own work (articles, poetry, art, music, video) in a public location that can be accessed by their classmates, friends, or anyone in the world. The web provides access to thousands of on-line magazines (ezines), newsletters, discussion groups, and entertainment sites that can broaden users' intellectual horizons (alas, some of them X-rated). It goes a long way to breaking down geographical, socioeconomic, cultural, gender, and age-related barriers toward accessing information and interacting with others.

Unlimited access to rich information resources brings with it numerous problems which educators and system administrators are trying to deal with. Since there is essentially no review process whatsoever for most information found on the web, the content of documents can be inaccurate, incorrect, or just plain fabrication. There is material which is obscene or in very poor taste. People are often overwhelmed by so much information, and they have no idea what to do with it all and how to weed through it. Finally, file structures and server addresses are often changed, which means that older links (more than a few days old) often do not work, resulting in wasted time and frustration. All of this leads up to the need for individuals to develop information-handling skills that allow them to cope with these issues.

Web-Based Workplace Training

Corporations have been somewhat slower to adopt and explore the web for learning. This is partly because, unlike academic settings, there is no tradition of Internet access, more limited availability of computers, and a concern about the confidential/proprietary nature of training materials. On the other hand, almost every corporation and business has a web site now — used primarily for marketing and product information. The percentage of employees with access to a computer in their workplace (not to mention at home) increases each year. Internal versions of the web (intranets), as well as external links to the Internet, are present or being developed in most organizations (Cronin; Gascoyne and Ozcubucku). So there is every reason to expect that the explosive growth of the web seen in the educational domain will be duplicated in the corporate world in the next few years.

How Digital Uses the Web for Field Training and Support

The 22,000 service technicians who belong to Digital Corporation's Multivendor Customer Services (MCS) group are responsible for installing and maintaining a wide range of computer systems. To train and support these technicians, Digital developed the MCS Learning Utility, an Internet/web-based performance support system which became available in 1995. Technicians access the network through one of 12 servers around the world (three in the U.S., six in Europe, and three in Asia). The network uses the web to provide access to internal and external databases that contain documentation and training materials including: hardware/software manuals, CBT courses from Digital and other vendors, information on available classroom training, and certification testing information. The MCS Learning Utility also has a search engine that allows technicians to search by part number, keywords, type of technology, content, or free text. It allows users to view courses and documents and then download them to their local system if needed. Using a laptop computer, a technician can access the system using a dial-in line from a customer site and obtain any technical information or training needed to complete the job. Says Rich Boucher, an MCS program manager, "The Learning Utility puts at the user's fingertips a single easy-to-use tool for accessing and retrieving a wide range of Digital and multivendor service information and training. The purpose of the [system] is to address the challenges of providing 'just-in-time' training and information for MCS service professionals around the world." The web provides an ideal way to do this for Digital as well as every other organization.

G. Benson and S. Cheney, (October, 1996) Best practices in training delivery, *Technical & Skills Training Magazine.*

On the hand, it should be acknowledged that many professional societies (particularly those in technical/technological fields such as engineering or telecommunications) have been fairly quick to put their publications and conference proceedings up on the web. To the extent that a significant amount of continuing education takes place through these organizations, we can say that the professional sector is taking advantage of the web for personal learning activities. In addition, many adults are involved in taking courses from post-secondary institutions and are using the web extensively. While the web may not be in formal use by many training departments at present, a large percentage of working individuals may already be engaged in web-based learning.

One of the most popular uses of the web is to provide customer education, either in the form of formal courses or more informal means such as newsletters and conference/chat areas. For example, the Microsoft On-line Institute (see Figures 14.1 and 14.2) provides a catalog of on-line classes related to their products, as well as a newsletter, conference, and career center. This site also serves as a showcase for Internet Explorer (a web browser marketed by Microsoft) in terms of the features used in the newsletter and conference facilitates. The IBM Global Education site (see Figure 14.3) also provides listing of courses relevant to its product offerings and services, as well as a directory of conferences/trade shows featuring IBM involvement, and listing of its consulting and certification programs. Visitors to the IBM site can select

Figure 14.1 The Main Web Page of the Microsoft On-line Institute Site

Figure 14.2 A Course Description from the Microsoft On-line Institute Site

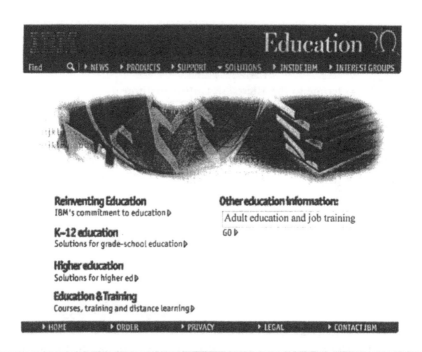

Figure 14.3   Features Provided in the IBM Global Services Web Site

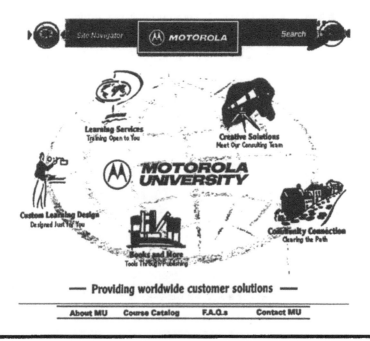

Figure 14.4   The Motorola University Web Site

the country or region of the world they are interested in seeing information about. Motorola University, the internal training component of Motorola Corporation, provides information about course offerings, career opportunities, and technical publications (print or electronic) at its web site (see Figure 14.4).

In addition to corporate sites offering information related to their products or services, a number of "virtual" training providers have appeared and are providing web-based courses, especially in the computer and technology domain. Given the worldwide nature of the audience and the minimum delivery costs involved, this is a very attractive marketplace. However, its growth will be constrained by the high costs of developing good multimedia courses and the availability of suitable hardware and high-speed network connections at customer locations. Developing a course that takes advantage of the latest multimedia features of the web but also can run successfully on a wide range of different customer systems is a difficult challenge.

The CyberTravel Agent

One of the most intriguing examples of web-based training to appear is the CyberTravel Specialist program developed by New Media Strategies and supported by the Institute of Certified Travel Agents and Hyatt Hotels (Figure 14.5). The program, which describes itself as the "Internet School for the Travel Professional," provides a full training course on how to use the Internet for travel-related activities. It consists of two components: a half-day lab session of supervised CBT conducted at CompUSA stores and a web-based course that can be taken anywhere. There is also an on-line certification test intended to verify that the student has achieved the competencies taught in the CBT component of the course.

This type of program could well be a model for future on-line training courses intended for independent professionals, featuring a certification test and a supervised lab session for those lacking computer background or when hands-on skills need to be taught. (See http://cybertravelspecialist.com).

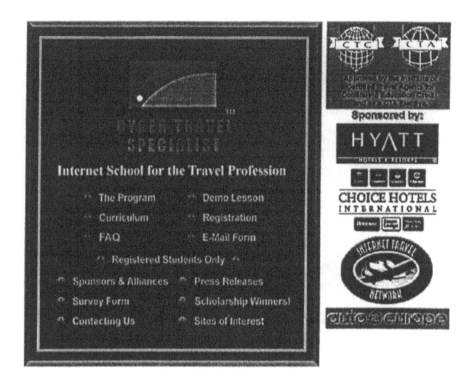

Figure 14.5   Screen Display from CyberTravel Program

LANs/WANs — the Network Workhorses

Within most organizations (including academic institutions), LANs are the form of network that people interact with most of the time. LANs are used for internal e-mail systems, shared access to databases and project documents, and all manner of budgeting/accounting activities. Much of the application software (e.g., word processing, page composition, spreadsheets, project management, CAD/CAM) and proprietary programs that employees use in their work run off LAN servers. In almost all organizations, the most pressing need to share information and data is with people in the same department or division — hence, a LAN is the most critical type of network.

Most academic institutions and a growing number of corporate training centers have computer labs in which all the machines are linked to a LAN server. This makes it possible to share use of programs and control access/software piracy. Furthermore, LANs make it possible to implement groupware and CSCW tools such as Lotus Notes, Microsoft Exchange, or Intel Proshare (Coleman). While such programs can run on any network, in order to obtain the immediate response time required, as well as ensure confidentiality of the data, a LAN is preferred. This is also true of programs that are very processing-intense, such as CAD/CAM, multimedia authoring systems, or mathematical modeling software. LANs are also fundamental to the electronic classrooms mentioned in the previous chapter.

Despite their overall importance, LANs tend to be the "invisible" technology in most organizations, getting little attention or mention from management. This is probably because their role is seen as secondary — to provide shared access to information and programs. However, LANs usually play a critical role in the transformation of an organization through information technology because they create the need for new work procedures and policies brought about by shared access. In almost all cases, shared access to documents, data, and programs changes the nature of peoples' jobs and their roles in some fashion.

A WAN (wide area network) is similar to a LAN except it is a network of computers that share the resources of one or more processors or servers over a relatively large geographic area. From an HRD perspective, WANs and LANs have very significant organizational impact that needs to be assessed and planned for.

Use of Lotus Notes at the Housing Development Board of Singapore

Lotus Notes is a widely used software program that demonstrates the power of LAN-based information sharing. The Housing Development Board of Singapore (HDB) is an example of an organization that uses Notes to increase its productivity, quality, and customer satisfaction. About 3000 employees of HDB use Notes in a Windows/Novell environment for e-mail, document databases, and staff/meeting scheduling. Some of the specific applications include:

- Press reply system — routes stories published in local newspapers about HDB or housing-related topics to appropriate departments and tracks replies
- Helpline system — routes all calls received though quality service employees to appropriate departments and tracks replies
- Staff information system — an on-line directory of all staff (including resumes)

- Parliamentary debates database — maintains extracts of all official parliamentary debates related to HDB and responses
- Medical benefits database — allows staff who are processing employee medical claims to identify appropriate medical benefits
- Conduct and discipline rules database — disseminates information about HDB conduct and discipline rules
- Sectional leave database — identifies leave schedules for key employees
- Staff suggestion database — employee suggestions for improvements

In addition to company-wide applications, some departments have developed Notes applications specific to their function. For example, the Contracts and Administration department has a variety of databases having to do with contracting and vendor activities.

The e-mail and database capabilities of Lotus Notes allows HDB employees to share information and manage workflow efficiently. However, the effectiveness of using Notes within the organization depends on user knowledge of the available applications and how to create new applications. End-user training is a limiting factor on the utility of this (or any other) organization-wide information technology.

L. Tung and E. Turban, (1997) Lotus Notes applications and impacts at the Housing Development Board (HDB) in Singapore. *Human Systems Management.* 16, 53-61.

It's a Wired World

While we have just emphasized the role of LANs and WANs in changing how an organization functions, any form of network could achieve this outcome (Martin; Sproul and Kiesler). This transformation caused by networking affects learning in a very broad and profound way. For one thing, it places attention on sharing resources and expertise. Since it is now much easier (or should be) to share ideas, experiences, and information, networks encourage collaboration and cooperation among employees. So the new form of learning in the networked organization is heavily dependent on electronic interaction via e-mail, conferencing, web sites, etc. Such interaction tends to be more informal and unstructured relative to traditional means of training courses and materials.

Knowing what and where information is available and how to access it becomes a very important skill in the networked organization. Part of this skill involves competence with a wide range of computer tools (i.e., the computer literacy discussed in Chapter 4). This requires constant retraining because hardware and software are continually changing. Another aspect of the skills needed are information-handling techniques; i.e., being able to weed through enormous amounts of data to find something useful. Use of search facilities and detailed knowledge of different information databases is important here. Finally, people need good communication and interpersonal skills, because much of the time they will be interacting with other human beings in search of information (albeit via electronic means).

The implications of networking for learning are even broader. Interacting with other people via electronic networks tends to break down all boundaries: geography, time, socioeconomic, cultural, age, and organizational. Individuals will seek out others with shared interests or common goals regardless of factors that would traditionally keep them apart or out-of-touch. This is the profound power of "cyberspace" or the "virtual community" that networks make possible (Jones; Rheingold). This means that employees can and will interact electronically with anyone in their organization as well as other organizations (perhaps even competitors). School kids communicate directly with university professors and scientists ... who talk to people in corporations and government agencies ... who can interact directly with consumers and their clientele.

So networks mean that learning is no longer encapsulated by artificial limitations such as classrooms, curricula, and organizational/institutional delimitations. Individuals will use networks to find other people and information sources that address their specific needs and interests. All organizations need to recognize this new circumstance and do what they can to enable their staff, students, or members to be as proficient as possible in the use of the networks.

Summary of Key Ideas about Networks

- The Internet is a network of networks, all of which use the same TCP/IP telecommunications protocol.
- The Internet provides for some common functions including e-mail, file transfer (FTP), listservs/news groups, and real-time chats (IRC).

- The world wide web (WWW) is an Internet application that allows file sharing by virtue of a common formatting language: HTM.
- Programs that can read HTML files are called "browsers," and multimedia components are handled by "plug-ins."
- Intranets are private versions of the Internet for use within a single organization.
- The web is widely used in K–12 and higher education, and is beginning to be used in training applications.
- Professional associations have been quick to develop web sites to provide member services.
- Corporations are using the web to provide customer information.
- Adoption and use of the web depends on the availability of suitable hardware and telecommunications connections.
- LANs are used in almost all organizations for electronic publishing, e-mail, and shared database applications.
- Some organizations are using LANs to support groupware activities, including decision support and design projects.
- Networks (LANs and WANs) encourage collaboration and interaction among employees and can change the nature of the information flow inside and outside organizations.
- Use of networks is an important employee skill in a wired world.

References

Berge, Z. and Collins, M. (1995) *Computer Mediated Communication and the On-line Classroom (Vols I-III)*. Cresskill, NJ: Hampton Press.

Campbell, D. and Campbell, M. (1995) *The Student's Guide to Doing Research on the Internet*. Reading, MA: Addison-Wesley.

Coleman, D. (1997) *Groupware: Collaborative Strategies for Corporate LANs and Intranets*. NJ: Prentice-Hall.

Cronin, M.J. (1996) *The Internet Strategy Handbook*. New York: McGraw-Hill.

Gascoyne, R.J. and Ozcubucku, K. (1996) *The Corporate Internet Planning Guide*. New York: Van Nostrand Reinhold.

Harasim, L., Hiltz, S.R., Teles, L., and Turoff, M. (1995) *Learning Networks: A Field Guide to Teaching and Learning On-line*. Cambridge, MA: MIT Press.

Jones, S.G. (1995) *Cybersociety*. Newbury Park, CA: Sage

Khan, B. (1997) *Web-based Instruction*. Englewood Cliffs, NJ: Educational Technology Publications.

Martin, J. (1996) *Cybercorp*. New York: AMACOM Books.

Rheingold, H. (1993) *The Virtual Community*. Reading, MA: Addison-Wesley.

Sproul, L. and Kiesler, S. (1991) *Connections: New Ways of Working in the Networked Organization*. Cambridge, MA: MIT Press.

15 | Knowledge Engineering — The Technological Future for Managing Knowledge

Douglas Weidner

A s organizations move toward better knowledge management, providing just-in-time learning, and becoming true learning organizations, knowledge engineering will become the *modus operandi* (Marquardt). In this chapter, we will describe the key elements of knowledge engineering, how knowledge engineering works via the technology of KnowledgeBases, and we will demonstrate how the practical use of these concepts can lead to a knowledge-based, learning organization. Finally, we will reflect upon these concepts and their implications for the future of knowledge engineering and the learning organization.

Foundations of Knowledge Engineering

The three key phases of business process reengineering (BPR) — development of strategic plan, modeling of "as-is," and design of "to-be" — illustrate the structured development of knowledge engineering as a discipline, its requirements, and the technology necessary to manage, and indeed leverage, knowledge. Proven BPR principles are valuable to guide the "radical redesign" of the entire concept of knowledge management for the organization.

Develop Strategic Plan

Knowledge has become the key corporate asset. Organizations obviously need to manage and leverage this key asset if they are to become a world-class organization. For this to occur, we need to radically reinvent our organizations so that they can truly manage knowledge and convert this knowledge into corporate success. Our vision, motivating our design, must be of a knowledge-based, learning organization benefiting the organization, its employees, and other stakeholders.

Model "As-Is"

Rather than modeling the entire organization to be redesigned, the conventional BPR approach, we need only describe the present state of the organization relative to the procedures, policies, and structures in which knowledge is managed. Table 15.1. describes the present, or "as-is," situation in many organizations relative to the method of training and the status of organizational knowledge.

Table 15.1 Present "As-Is" Description of Many Existing Organizations

Topic	Status	Problem or Gap
Training	Provided in formal classes	The nature of training is "Just-in-Case" training — everything I may need to know at some future, unspecified date instead of what I need to know now to get this specific task done.
Procedures	Documented in procedure manuals	Obsolete or poorly maintained and fallen from use.
Organization Data and Information	Warehoused	Data is plentiful and organized, but how to use it effectively is not so obvious or well documented.
Knowledge workers	Substantial portion of work force	They are often asked to do things outside their customary tasks and scope of past experience.
Knowledge	Knowledge needs are well documented and often talked about	Little has been done to manage or leverage knowledge.

Let's personalize the "as-is" description. Have you ever been a typical white collar, knowledge worker? You probably found yourself running here for the tools you needed, there to get written or verbal guidance on what to do, a third place to get accurate data or information, if it was available at all. Finally, you might have had to guess at what your deliverable, or product, should look like. Unfortunately, this is the typical knowledge worker scenario and dilemma.

Knowledge is power, but instead of feeling empowered, most of us feel overwhelmed by the vast resources available to us. A dizzying array of publications, procedures manuals, memos, staff meetings, training classes, videos, and other technology devices can be especially overwhelming in a large corporation or an organization such as a government agency.

The present paradigm expects employees to capitalize on knowledge from such a fragmented collection of resources. This depressing situation exists all too often, despite our technological innovations, information technology investments, human resource emphases, new organizational designs, process re-engineering, and continuous process improvements.

Design "To-Be"

One of the key components of this step is to envision the future. There are many tools and techniques to aid this process. For example, one might add a fourth column to Table 15.1 called "vision" and describe for each factor what might be envisioned as a radical solution or what a new way of doing business might look like. The sum of all these visions might begin to focus creativity on a new process (system and way of doing business) that could successfully enable a substantial number, if not all, of the individual visions.

Another approach might be to research paradigms that exist in other disciplines or industries, or possibly in other functions within your organization or among your benchmarking partners. You could test to see if these external paradigms might represent a radical, but appropriate, new way of doing business in the organization being reviewed. In other words, the key asset of the learning organization is knowledge. Can we find another paradigm that could be translated into our knowledge-management environment?

For instance, consider the cooking profession. Have you ever watched a professional chef? All the utensils, recipe instructions, and ingredients are readied before cooking begins. This practice, called *mise en place* by cooking professionals, is what separates the pros from the mere players, the gourmet meal from the TV dinner. Your expectation is excellence and it is exceeded. Can we translate the chef paradigm into something meaningful regarding knowledge?

The gap between these two scenarios can give us many insights into a new paradigm that describes how an agile, learning organization should function and the benefits it should reap. Compare the typical knowledge-worker scenario described above to the chef paradigm. The knowledge-worker scurries around to find the knowledge necessary to perform the task at hand; the professional chef has all ingredients in place before commencing.

Try this vision on for size. It is one possibility of how the *mise en place* practice or chef metaphor translates into the *modus operandi* for the knowledge-based, learning organization.

> Learning Organization Vision: "Provide the best knowledge (ingredients and recipe) to the right person (chef) just in time or at the right time (when an order is placed)."

The next step in this scientific approach to designing a knowledge management approach is to translate the vision into a hypothesis to be tested for validity. Typically, a hypothesis is a theory that can be tested against existing and proven laws or with empirical research, such as surveys or controlled clinical experiments.

Axioms, Corollaries, and Architectural Layers

For our purposes here, two issues are worth noting: 1) let's be a little less rigorous and propose an hypothesis made up of *axioms* (a proposition regarded as a self-evident truth), e.g., "best knowledge," "right person," etc., and *corollaries* (an immediate inference from a proved proposition, something that naturally follows), e.g., best knowledge must be "relevant to the task at hand," and then, 2) let's simultaneously address *three architectural layers* of a potential solution — namely, the knowledge itself, the functionality or structure of the knowledge (base) that serves as a repository, and the knowledge delivery or access mechanism. This is a complex exercise, but the process should promote reflection and understanding about key factors necessary for knowledge management and the leveraging of knowledge. Table 15.2 outlines some implications of just a few key axioms and corollaries of the "to-be" design hypothesis on the three architectural levels of knowledge, KnowledgeBase functionality, and delivery mechanism.

Table 15.2 Knowledge Axioms and Corollaries on Three Architectural Layers

Hypotheses	Architectural Layers		
	Knowledge	*KnowledgeBase Functionality*	*Knowledge Delivery/Access Mechanisms*
Axiom 1: Best Knowledge			
Corollaries			
Relevant knowledge only	Process/task oriented	Process-oriented search	Access to software tools
Threshold of amount	Beneficial even if partial knowledge	knowledge building blocks/feedback compatible	Economical at any level
Sound knowledge/instructional design	Abides by proven knowledge theory	Supports sound theoretical construct	Delivers sound theoretical construct
Axiom 2: Right Person			
Corollaries			
Knowledge seeker	Need for multiple perspectives/viewpoints	Accommodates multiple viewpoints	Universal access
User compatibility	Diversity of user learning Needs	Accommodates user diversity	Multimedia capable
Axiom 3: Right Time			
Corollaries			
Existing knowledge	Explicit knowledge vs. tacit	Previously structured knowledge	Universal and easy accessibility — user
Age of knowledge	Updated frequently	Easy to author/enrich	Update speed, accessibility to author

Just as the chef metaphor — having all the ingredients ready beforehand — provided insight into an appropriate metaphor or paradigm for knowledge, "the best knowledge to the right person just in time," the reverse process of combining the requirements or architectural components can result in the needed knowledge management system design. Let's expand and clarify each of the three new paradigms.

Best Knowledge

In the old paradigm, the burden of search was on the knowledge seeker. In the new paradigm, the appropriate ingredients are arranged before the chef in an organized manner at the moment of need. In the business world, typically, the relevant knowledge is the knowledge associated with the performance of a specific activity or task within a known process. The process then becomes a more powerful organizing scheme than the most robust search algorithm applied to general knowledge. Other issues, that could be developed more fully, include: the threshold amount of knowledge needed and whether the provided knowledge needs to abide by sound instructional design principles, which of course it should.

Right Person

In the old paradigm, a search algorithm applied to a database might yield textual material written from the perspective of a technical analyst when the seeker might need a management approach or perspective. This need for multiple viewpoints and perspectives, as well as the necessary attention to diverse learning needs, is critical to system design if the system is to be truly effective.

Right Time

In the old paradigm, knowledge is sought when needed. Often this necessitates the conversion of unstructured tacit knowledge, in the heads of subject matter experts, into explicit usable knowledge. This has proven to be a difficult, if not impossible, just-in-time exercise. In the new paradigm, the ingredients or necessary knowledge must be amassed beforehand, knowing the need associated with common processes. But even with knowledge gathered beforehand, ease of access and use is not guaranteed. In addition, ease of use is as important for the author as for the user if the knowledge is to be maintained and enriched.

From Food for the Stomach to Food for the Organization

Let's continue the chef metaphor to define the critical components of such a knowledge management system. The knowledge management system is the seamless integration of: 1) a robust, proven, process methodology or set of procedures that serve as the organizing scheme for the knowledge, 2) a broad array of commercial off-the-shelf (COTS) software tools that enable accomplishment of the task at hand, and 3) just-in-time (JIT) knowledge and training, which supports each task by synchronizing it with the appropriate knowledge references and JIT training. Specifically, such a knowledge management system includes all the components that might describe a professional chef, including:

Recipe Instructions — The knowledge management system defines each step-by-step task in the methodology or business process. In addition, JIT knowledge and training are provided to empower employees, to provide a basis for intuitive knowledge to be gained through experience. The knowledge management system provides knowledge, not just information. In this sense, knowledge is information that has value, e.g., knowledge is what to do with information. Its value is its accessibility at the right time and place, ready to be applied. Its application and positive impact reinforces intuition and learning.

Utensils — The knowledge management system links each task with the best COTS tool available to accomplish the task, even preformatting the tool for the specific task at hand, if feasible.

Ingredients — The knowledge management system supports each task by synchronizing it with the appropriate inputs (data/information) and controls (guidance), while specifying the needed resources in terms of people, equipment, and facilities.

The Meal Presentation — The knowledge management system prevents guessing about what the final product should look like by providing a repository of past deliverables (gourmet feasts). A key collaborative feature: pointers to others within the organization, who are willing and able to provide assistance to the new cook on his/her way to becoming an empowered chef, are provided.

Knowledge engineering is more than the combination of work flow management, an electronic performance support system, the learning organization, or any specific business discipline such as Operations Research, Total Quality Management, or Business Process Reengineering. It clearly has its roots in all of these important management and learning concepts and disciplines, but enhances them by a simple, but highly critical paradigm shift.

Knowledge engineering helps the chef cook a sumptuous corporate feast, after which no organization will accept junk food again.

Introducing the KnowledgeBase Tool

So far we have been talking theory and paradigms. Let's now convert theory into reality and practice. We will look at the necessary steps, the organizational structure of a knowledge repository, and the tools that create a knowledge management technology called KnowledgeBase Tool, originally designed for the U.S. Department of Defense during the period 1994–1996 as *Process Management Tool* (Weidner, 1995). Before looking at KnowledgeBase Tool, it is important to explore its essence, i.e., a KnowledgeBase.

KnowledgeBase

A KnowledgeBase consists of two primary types of synchronized information, a work breakdown structure (WBS) and references:

Work Breakdown Structure

The WBS is a template of activities and tasks to be performed in order to complete the process or project, the "who," "what," and "when," for each task (Figure 15.1). Technically, a WBS is a process decomposition model which is a fully documented activity node tree of activities and tasks. The WBS is the backbone or skeleton of the KnowledgeBase or the organizing scheme for the process knowledge, which is included in References. The WBS should be exportable to project management software for project control purposes, e.g., a Gantt Chart. The process documentation should include the inputs and outputs for each process as well as the controls (constraints) limiting each activity and the mechanisms, the people, equipment, and facilities involved in each activity of the process.

Knowledgebase Components

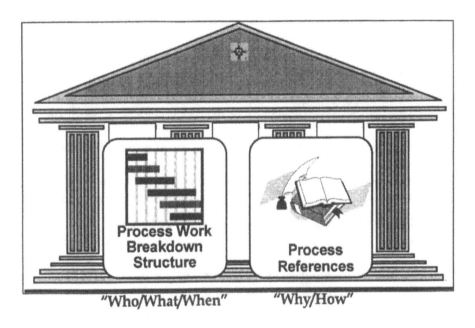

Figure 15.1 Two KnowledgeBase Components — WBS and References

References

References are the meat or muscle of the KnowledgeBase. They provide the process knowledge — lessons, guidelines, tools, and techniques. The more synchronized references provide the process details and JIT training to support each and every task, the "why" and "how" for each task. There are various types of references including educational, training, practical/operational techniques, work products, and resources. (See Table 15.3 for definitions of references by applicable user, i.e., the intellectual content of the KnowledgeBase.)

Work Breakdown Structure of KnowledgeBase Tool

Figure 15.2 shows the first of the two primary KnowledgeBase Tool screens corresponding to the two KnowledgeBase components. This is the Work Breakdown Structure (WBS). The following is a brief description of some of

Table 15.3   Definitions of References for Users of KnowledgeBase

Educational Background — Project Manager and Team Members
  Lessons — Prerequisite preparation — Background educational and instructional materials, including word documents, graphic presentations, even computer-based-training or interactive video/multimedia
  Bibliography — Additional resources — Books/recent periodicals (author, source, and summary/abstract)

Educational/Training Concepts — Project Manager and Team Members
  Guidelines — Overview of pertinent methodology — methods, tools, and techniques
  Keys to Success — Summary of recommendations — increase likelihood of success
  Barriers — Summary of roadblocks or obstacles one might face
  Checklists — List of items to review prior to commencement, e.g., prerequisites

Practical/Operational Specifics — Typically, Team Members
  Techniques — Technical methods to be used — formulas, notation, examples, etc.
  Tools — Link to systems that automate a specified method or technique
  Steps — Sequence of actions to execute a method, technique, or task

Work Products (Inputs/Outputs/Controls) — Team Members
  Templates — Preformatted documents and examples — Word, Excel, PPT, etc.
  Formal Documents — Published documents — typically corporate policies and procedures.
  Informal Documents — Repository — prior work products or research notes.

Resources (Mechanisms) — Project Manager
  Personnel — Required staff/qualifications needed to perform the Activity or Task.
  Facilities — Required physical assets needed to perform the Activity or Task.
  Equipment — Required tools and supplies needed to perform the Activity or Task.
  Other — Miscellaneous — one or more additional reference categories such as "Frequently Asked Questions"

the KnowledgeBase Tool functionality associated with the WBS (Figure 15.2). The WBS is the backbone or skeleton of the KnowledgeBase, the organizing scheme for the process knowledge and references.

The screen in Figure 15.2 shows the WBS on the left side. For each WBS activity (in this case A412 Model As-Is), the label, title, and description are shown on the right-hand side of the screen. This option is noted in the

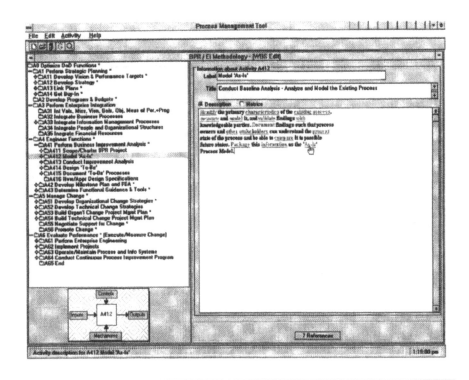

Figure 15.2 KnowledgeBase Tool Depiction of a Process WBS

context sensitive prompt in the lower left-hand corner. A powerful feature of this WBS representation is the ability to always maintain positional awareness — where you are in the overall KnowledgeBase methodology. Sometimes the worker needs clarification/definition of words or acronyms used in a description. Key words in the description text are highlighted, in this case "as-is." They are defined by merely clicking on them. This functionality enables immediate and better understanding of text descriptions.

Activity Tasks in KnowledgeBase Tool

Figure 15.3 once again shows the same KnowledgeBase Tool screen, but this time we have further decomposed the methodology to uncover Tasks. Each terminal or leaf activity is comprised of "Tasks," the elemental work units comprising the activity. Knowledge of these Tasks is essential to actual accomplishment of the activity.

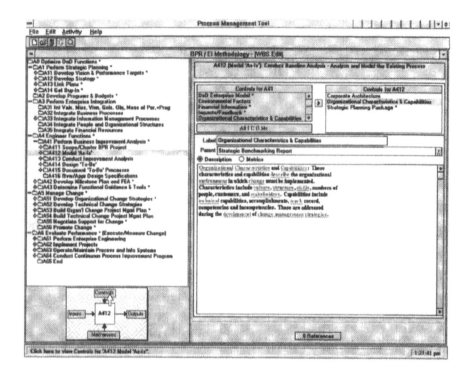

Figure 15.3   KnowledgeBase Tool Depiction of Activity Tasks

1. We have further decomposed the A413 activity, "Conduct Improvement Analysis." This is done by clicking on the plus (+) signs in front of an activity such as A413 until we reach leaf or terminal activities such as A4133 "Identify Performance Gaps." When an activity has been decomposed, the plus sign changes to a minus sign, which now is displayed in front of A413.

2. For leaf activities (terminal activities with no further decomposition), a Tasks object appears in the IDEF icon in the lower left-hand corner. Clicking on the Tasks object enables viewing of an activity's tasks, as noted in the context-sensitive prompt below the arrow.

3. The six tasks comprising activity A4133 are listed on the right-hand side of the screen.

4. Selecting a task (2. "Document Gaps — Process Cycle Time") displays its description on the lower, right-hand side of the same WBS screen.

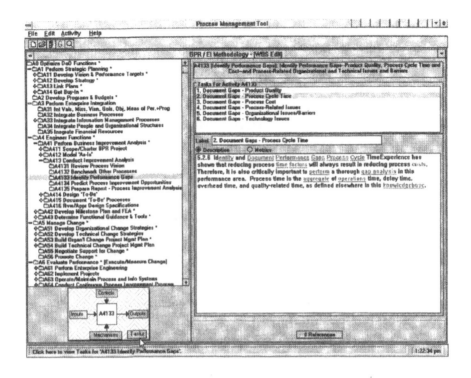

Figure 15.4   KnowledgeBase Tool Depiction of References — The Books of Knowledge

One of the key design strengths of the KnowledgeBase Tool is the ability to always maintain knowledgebase context or positional reference no matter how deep into the KnowledgeBase we dig. That is, the full KnowledgeBase methodology is always available on the left-hand side of the screen.

References for KnowledgeBase Tool

Figure 15.4 shows the contents of the second major KnowledgeBase Tool screen, 'References'. This bookshelf-like screen is accessed by clicking on the Reference button in the lower right-hand corner of the former WBS screen, for any activity, task, or ICOM, as described above. Knowledge of References is essential to efficient and effective accomplishment of the activity. Ultimately, References becomes the feedback mechanism for continuous process improvement.

References are the muscle of the KnowledgeBase. They provide the details: JIT training, techniques (guidelines, keys to success, checklists, etc.), and tools to support each and every task (as depicted in Table 15.3). References are the "why" and "how it must be done" for each task. Under the new paradigm, references are included in a *unique* Reference Bookshelf for each activity, i.e., only knowledge related to that activity is accessible at that point in the search. The Knowledge Engineer ensures that relevant knowledge is correctly referenced, so that the user does not need to search through reams of paper and electronic materials, wasting time tracking endlessly into hypertext cul-de-sacs.

1. The Reference screen provides a bookshelf across the top of the screen, the contents of which are displayed — type of Reference Book and Reference Title — on the left-hand side of the screen. A user has two choices:

   - Click on one of the books itself to see its entire contents. This is a desirable feature when the list is long and extends beyond the bottom of the window.
   - Select a book from the list, such as has been done above (Lessons — "Case Study in Change Management").

   The reference library is *unique*; it contains only knowledge related to *this activity.* Hence, the KnowledgeBase contains potentially hundreds of unique reference bookshelves, one for each activity and task. The advantage of this innovative approach is that the user does not have to comb through reams of information or seemingly endless hypertext links to find applicable useful kernels of knowledge. That search has been done by the Knowledge Engineers who created the KnowledgeBase.

2. Each book entry has a memo field and may have an object. The memo field, on the right-hand side of the screen, provides the content of that selection from the chosen book. If an object exists, the memo field introduces the object. An object is used when the text-based memo field is inadequate to handle the stored knowledge because of size or format requirements, such as use of graphics or multimedia.

3. In this example, the user can access the object merely by clicking on the Read button.

4. The 'Read' button launches the object, in this case a specific lesson that addresses a case study that may be of interest. Once the object is launched, any of its contents can be copied to other documents, if

needed. The advantage here is that KnowledgeBase contents can be used to help fulfill project requirements.

The objects can be anything Object Linking and Embedding protocol (OLE) can support, including Word documents as templates, tools to accomplish the task, or multimedia for just-in-time training.

Benefits of Using a KnowledgeBase Tool

By using the KnowledgeBase Tool and resident process KnowledgeBase, the team will substantially increase project efficiency and effectiveness, create better and faster results, and be able to continuously improve the process. The KnowledgeBase Tool meets these needs by addressing two significant ways project outcomes can be improved: a) better project control, and b) better informed project team members. The KnowledgeBase Tool is the front-end user interface, an easy-to-use knowledge organizer that makes a seamless, end-to-end proven methodology even more practical and functional.

Project Management and Knowledge Repository

The KnowledgeBase tool can serve both as a project management tool and a knowledge base repository. Project management functionality has been designed into the KnowledgeBase Tool, in accordance with the axiom which states that projects should be scheduled using Gantt or PERT charting techniques to develop, track, and display scheduling data by task. The KnowledgeBase Tool facilitates this activity and ensures consistently high-quality results.

Using the KnowledgeBase Tool, the project manager exports the WBS: 1) from the KnowledgeBase Tool directly to COTS project management software (e.g., MS Project) which includes project control functionality, especially Gantt and PERT charts; 2) the project manager then selects the applicable activities (or deselects activities that are not necessary) for the specific task at hand. In MS Project, the project manager can fine-tune parameters for planned duration, timing, resource assignment, etc.

The process KnowledgeBase provides a comprehensive superset of project management tasks, ensuring that nothing is forgotten or overlooked. It eliminates the time-consuming labor of developing project plans from scratch, reduces the error-prone tinkering that was formerly inevitable, encourages reuse of previously successful work plans, and promotes consensus by utilizing authoritative resources. By exporting directly to MS Project, the Knowledge-Base reduces errors and facilitates effective project control.

Team members use the KnowledgeBase Tool as a quick, convenient source for lessons, guidelines, keys to success, best practices, tools, templates, etc. The KnowledgeBase Tool Reference window provides ready access to the KnowledgeBase references. The close link between the KnowledgeBase and the process steps exported to MS Project as a WBS ensures that no matter how many activities are excluded from the WBS, the available references always correspond to the reduced set. This means that project team members have immediate access to the specific knowledge directly related to their assigned task. Having this knowledge at their fingertips substantially reduces the inefficient searching for knowledge and ensures that project team members are properly prepared to perform their tasks.

Two powerful results (better costs and better results) are obtained by using the knowledge-based approach based on integrating your process as a Knowl-edgeBase with the KnowledgeBase Tool.

Project Performance (Fast/Low Cost)

The project team improves project efficiency (project completed on time and on budget) based on better management and control; and the project team improves project effectiveness (implementable with existing and emerging technologies) based on knowledge-empowered team members.

Process Optimization (Better Results)

The project team debriefs its efforts and continuously enhances the process KnowledgeBase with lessons learned, including project Metrics, Tem-plates/Deliverables, and emerging Tools that may have been tested and which can aid future projects.

Finally, as an added benefit, the use of a process KnowledgeBase instills a knowledge-based, learning organization orientation into project team members. For those process improvement projects, their To-Be designs will be influenced by their use of leading-edge knowledge-based tools, i.e., their design will benefit from some emerging, powerful concepts of being a knowl-edge-based, learning organization.

Using KnowledgeBases to Create a Learning Organization

Knowledge engineering has defined two different primary uses for KnowledgeBases and the KnowledgeBase Tool functionality. KnowledgeBases can be used effectively: 1) for designing an entirely new process, the "to-be"; and 2) for the ongoing optimization of an existing process. In both cases, the Knowledge Engineering principles as demonstrated through the KnowledgeBase Tool functionality fosters the learning organization.

Designing a New Process, the "To-Be"

Creating viable "to-be" scenarios is easier said than done. Many BPR teams have been so heavily immersed in modeling "as-is" processes that they have had little experience at envisioning the future. Also, their analytical and technological skills often ignored the people side of change. They have designed the "to-be" using state-of-the-art technology, giving lip service to the fundamental requirements of organizational change management, but little substance. They have failed to recognize that technology insertion is doomed to fail unless the underlying processes are also changed, since people will evolve their own responses to technology, often in a haphazard, nonoptimal fashion. The KnowledgeBase Tool can provide needed structure to the "to-be" process design, while self-generating training, procedural, and reference materials.

The team leader or project manager designates a project Knowledge Engineer who creates a "to-be" Process KnowledgeBase on the KnowledgeBase Tool concurrently with the process design effort. Procedures are documented and included in the "to-be" Process KnowledgeBase by the Knowledge Engineer, essentially creating a procedures manual as they go, with each procedure instantly available to every member of the Team. With respect to training, bite-sized lessons can be created and added incrementally to the "to-be" Process KnowledgeBase as they are completed.

How do the outcomes compare? In the traditional approach, actual start-up day can be quite taxing, especially when training and procedures manuals have yet to be published. The re-engineering team spends precious time floundering around, bemoaning the bureaucratic obstacles, while their initial enthusiasm and motivation fades. The new, knowledge-based approach using the KnowledgeBase Tool and "to-be" Process KnowledgeBase, enables the team to hit the ground running on day one. Such an approach makes all the difference, yet the cost is the same. The benefit comes during the job from managing and improving the implementation process, not spending more money.

Optimizing Existing Processes

Thus far we have discussed how a knowledge-based approach can smooth the new process start-up phase. Let us now briefly summarize how the same knowledge-based approach can optimize the new process through continuous process improvement. Over time the "to-be" Process user will begin to establish best practices. These best practices will come from lessons learned, noted keys to success, and accumulated metrics. Such knowledge, properly used, enriches the "to-be" Process and completes the feedback loop. Such a program needs to be designed into the process from the outset.

The problem is that these insights and collected metrics need a repository and a way of sharing this knowledge to become a knowledge-based learning organization. Process KnowledgeBases, resident in the KnowledgeBase Tool, are just such a vehicle for sharing. They provide structured knowledge on a just-in-time basis, and enable knowledge sharing and the associated cultural benefits.

How to Create a KnowledgeBase

Unlike a procedure manual, a KnowledgeBase is a living repository. It may originate with a procedure manual or an accepted methodology, but it grows and becomes enriched by capturing lessons learned, examples of prior art/architectures, and a more inclusive tool set, as well as by fostering knowledge accumulation, sharing, and growth.

A Knowledge Engineer creates an initial KnowledgeBase by one of two primary methods: by population of the KnowledgeBase using existing documented knowledge, or by conversion of tacit knowledge, in the heads of subject matter experts, into explicit, structured knowledge. Typically, both methods are used, but here is a brief explanation of each.

Documented Knowledge

A considerable body of knowledge exists for most business processes. This knowledge is in the form of procedures manuals, training courses, forms, checklists, policies, etc. In this first case, the Knowledge Engineer's task is to collect the existing resources, build a work breakdown structure that represents the process, and categorize the knowledge according to the process' activities and tasks. In addition to the knowledge researching activity, process modeling skills are often required. This case may result in sufficient knowledge to build the initial KnowledgeBase, or these activities may serve as the springboard for the second major type of knowledge collection.

Undocumented, Tacit Knowledge

This type of knowledge is more difficult to collect. Traditional means include the use of interview techniques on various subject matter experts. A more powerful approach uses group decision support software, often called groupware, to both elicit knowledge from experts and reach consensus among these experts, often in a single session.

For both methods, good instructional design skills are needed. Whether the knowledge is initially documented or undocumented, the result is the same — an initial KnowledgeBase that will be enriched over time.

How to Use the KnowledgeBase

Various knowledge workers use the KnowledgeBase to improve work performance, to accomplish the defined process. The following are some examples of KnowledgeBase use assuming basic unfamiliarity of this specific process by the knowledge worker.

Initial Use

The first time a knowledge worker A (KW-A) uses the KnowledgeBase, it is likely that a "Lesson" is appropriate. Rather than "just-in-case" training delivered months beforehand with low retention levels, the "Lesson" is delivered just-in-time to meet the need. Once trained, KW-A can take advantage of other "Reference" materials as needed, including "Tools," "Guidelines," "Keys to Success," etc. Let's consider a few other possible events leading to task accomplishment.

Policies

In the old paradigm, where the burden of acquiring the needed knowledge is on KW-A, any relevant policies are documented in policy manuals and it is KW-A's responsibility to comply with them. Often such compliance is bypassed due to time pressures. But this is the new knowledge paradigm where the chef's ingredients are all available. In the new KnowledgeBase environment, the Knowledge Engineer has researched applicable policies; they might be listed in "Formal Documents," not as complete electronic versions of the policy manual, but the actual, applicable chapter and verse, with a clarification if necessary. In this way, KW-A knows what policy applies, and its applicability is even amplified if deemed unclear by the Knowledge Engineer.

Past Work Product Examples

At this stage of the KnowledgeBase development, probably no examples of past work products exist. KW-A may be on his/her own regarding the look and feel of the final product. This situation will improve as the Knowledge-Base is enriched.

Tools

If no "Tool" was recommended for use, KW-A must search for a tool to aid completion of the task at hand, possibly using a spreadsheet for analysis.

Even with training, KW-A's task is not trivial. Now let's see how performance can improve over time.

Subsequent Use

The next time KW-A comes down this path in the KnowledgeBase, i.e., is required to perform this activity, the situation has improved. Here's how:

"Guidelines" and "Keys to Success"

If KW-A read these "References" without the benefit of the context for these recommendations, they would be meaningless. Because KW-A took the "Lesson," even if months earlier, these references trigger recollection of the "Lesson," content. The "Guidelines" and "Keys to Success," if they exist, become meaningful because the "Lesson" was taken. They become the key review material, obviating the need to retake the lesson.

Past Work Product Examples

By now, a few other knowledge workers, B, C, D, etc., have created work products. These are available for KW-A to reference. Often, these examples may go a long way toward KW-A's accomplishment of the task.

Tools

KW-A need not search for a tool to aid completion of the task at hand. The one used last time is available.

Enriching and Improving the Knowledgebase

The Knowledge Engineer continuously enriches the KnowledgeBase with new tools that have proven to be effective, past deliverables that might serve as examples for future tasks, and experiences or lessons learned in past efforts. The Knowledge Engineer focuses this new knowledge and further populates or enriches the KnowledgeBase with it. The KnowledgeBase is enriched by adding, for example: 1) further, clarifying decompositions of existing activities; 2) new references in the Knowledge Bookshelf, such as policies that may apply; or new "Lessons" that need to be created and added to respond to user needs, either on methodology or specific tool use. Here are some examples of how this works.

KnowledgeBase Debriefing

It is the joint responsibility of the Knowledge Engineer and knowledge worker users of the KnowledgeBase to enrich it. In a debriefing session the Knowledge Engineer guides the past users toward documentation of the following:

1. *Past Deliverables* — The Knowledge Engineer collects past deliverables for joint analysis. Sometimes a consensus can be reached, taking the best from each to create a template that best satisfies future needs. Consider how much more efficient it is to work from a template than to treat each new attempt in an *ad hoc* fashion.
2. *Tools* — Each past user may have chosen a different tool. The best can be decided upon and made universally available. It is even possible

that a worksheet can be loaded as the starting point for future users. Consider how much more efficient it is to have the recommended tool at your fingertips, just when you need it.

3. *Lessons Learned* — This is the area of true wisdom and understanding. The lessons learned need to be included in "Guidelines" and "Keys to Success." Sometimes the lessons learned lead to complete revision of the process. If this sounds like the essence of continuous process improvement and TQM, it is.

4. *KnowledgeBase Shortcomings* — The Knowledge Engineer's true success is not in the initial quality of the KnowledgeBase, but in its ultimate quality and usefulness. Feedback concerning the inadequacy of "Lessons," frequently asked questions, etc., all promote improvement. For some tasks, frequently asked questions can be listed and answered in the form of a new "Reference Book." As an extra benefit, the e-mail or voice mail address of an expert can be listed for those occasions when the question has not been addressed in frequently asked questions.

Future of Knowledge Engineering

There are four requirements for Knowledge Engineering to successfully result in a true learning organization: 1) top management commitment, 2) employee involvement, 3) dedicated and capable knowledge engineers, and 4) appropriate technology. As with many other process improvement disciplines and initiatives, the basic technology is readily available. Here are some of the other considerations:

1. Top Management Commitment — A necessary but not sufficient requirement for success. Top management vision and leadership is an essential ingredient (Weidner, 1997). This axiom has been addressed by many business researchers because it has proven to be the downfall of many past initiatives including TQM and BPR.

2. *Employee Involvement* — The adage, "You can lead a horse to water, but you can't make it drink" is appropriate. Much future research in the knowledge management arena must be aimed at this critical link in the chain. Employees, above all, must believe they will benefit from KnowledgeBase use; such use must become second nature; and, they must be empowered to contribute to, as well as be rewarded for contributions to KnowledgeBase development. Employee contributions to process

improvement and total quality management have been well documented in the literature. Here is the new challenge: we must engineer their involvement with knowledge acquisition and enrichment!

3. *Capable Knowledge Engineers* — A new class of knowledge workers must evolve from the ranks of trainers, procedure manual authors, and process improvement professionals. These new leaders must be capable instructional designers, process modelers, and motivators.

4. *Technology* — The concept of KnowledgeBases has been proven using existing technology — PCs and LANs. The immediate future will include the need to make KnowledgeBases even more universally available by capitalizing on the ubiquitousness of Internet capabilities.

Knowledge Engineering and the KnowledgeBase approach represent the future of managing knowledge in organizations. They provide the hope to enable companies to leverage knowledge and to become learning organizations. The benefits are so powerful, intuitive, and elegant, that most of us will wonder why we did not "just do it." A few others will prefer to remain with the old paradigm. With it, the burden to assemble all the relevant knowledge, tools, and historic repositories remains with the user, a "bus-boy" mentality, rather than the chef's *mise en place* by the subject-matter experts.

Summary of Key Ideas
About Knowledge Engineering

- Knowledge engineering is built on principles and lessons of business process re-engineering (BPR).
- The three key phases of BPR are development of strategic plan, modeling of "as-is," and design of "to-be."
- The architectural layers of knowledge engineering include best knowledge, right person, and right time.
- KnowledgeBases consist of work breakdown structure (WBS) and references.
- KnowledgeBase Tool is a comprehensive EPSS that assists and guides workers through their various tasks.
- Workers can use KnowledgeBase Tool for lessons, guidelines, keys to success, best practices, tools, templates, etc.
- KnowledgeBases are valuable in creating a learning organization.

References

Marquardt, M. (1996) *Building the Learning Organization — A Systems Approach to Quantum Improvement and Global Success.* New York: McGraw-Hill.

Weidner, D. (February, 1995) An Interactive Enterprise Integration/Business Process Reengineering Empowerment Tool for DoD. Proceedings of Third International Symposium on Productivity & Quality Improvement With a Focus on Government. Washington, D.C.

Weidner, D. (1997) *What Senior Managers Need to Know About BPR.* Unpublished. (Available from the author; see below.)

About the Author

Douglas Weidner, Chief Knowledge Engineer with Litton/PRC's Defense Services Management Consulting Group, is at the forefront of knowledge engineering, having designed the *Process Management Tool* and developed the first KnowledgeBase for the U.S. Department of Defense. He is an engineering graduate of the U.S. Air Force Academy with an MBA in Business Economics and an MSIE in Operations Research, and is presently working on his Ph.D., specializing in Executive Leadership through Knowledge.

16 Selecting and Evaluating Technology

This chapter examines the issues associated with selecting the most effective and cost/benefit technologies for learning and knowledge management. We will also consider quality control and assessment as they pertain to technology. Each of these topics is a different aspect of evaluation and fundamental to achieving performance improvement and corporate success.

Technology for Learning and Managing Knowledge: Does It Work?

Without a doubt, the most popular question about the use of technology is whether or not it works! There are a number of ways this question can be interpreted, but the most common version is whether technology-based learning is as effective as traditional classroom instruction. The answer is a clear-cut "yes" by any measure.

People have been conducting studies to compare a particular training course taught in the traditional (classroom) manner with one delivered by some form of technology for decades (Kirkpatrick), and the results are almost always the same — no significant differences in terms of learning outcomes. In other words, most studies show that you get the same results whether you use technology or not. So we can say that technology-based approaches work just as well as conventional classroom instruction. Furthermore, trainees typically report that they find technology-based training as satisfying as classroom experiences — and in some cases, much better.

There is more to the story, however. The same studies also show that technology-based training takes much less time to complete relative to classroom instruction (on the order of 25 to 50% less). This is a very significant finding because it means that training can be delivered at a lower cost if less time is needed. And because it takes less time to complete, employees may prefer it. Interestingly, the reason it takes less time is not strictly a function of technology use, but a consequence of better instructional design (making information easier to understand and getting rid of irrelevant material) as well as the opportunity for individualized learning, which allows people to progress at their own pace. But improved design and individualized presentations tend to be an integral aspect of most technology approaches.

There are a few notable exceptions to the "no significant difference" phenomenon. One domain where learning improvements have been well documented is the use of simulators/simulations. In particular, the U.S. Department of Defense has conducted numerous studies showing that job/task performance is improved after simulator training (in addition to training time reductions). Studies also appear to show that many interactive multimedia programs result in better learning in terms of increased retention, better comprehension, or transfer of skills/knowledge — relative to comparable classroom learning experiences. Of course, this is a difficult comparison to make since a well done multimedia program is nothing at all like sitting in a corporate training room listening to a boring lecture!

This last point does underscore a very important consideration about such comparison studies — the assumption that they involve the same kind of learning activities or processes — which they probably don't. Learning something from a classroom lecture or training meeting is not at all the same as learning the same material through an interactive CD-ROM, a television program, or a two-way videoconference. Each medium and delivery system has its respective strengths and weaknesses, and it seems unlikely that people acquire the same skills/knowledge from different approaches. On the other hand, if learners are properly motivated, they will be able to learn what they need/want to know from any media.

Cost/Benefits Analysis

While cost/benefit techniques ought to be practiced in any form of training endeavor, they become especially important in technology applications because the magnitude of the financial investment involved is substantial (equivalent to those made in other major business sectors, such as manufacturing, marketing, or product development). Furthermore, technology applications typically involve innovations and significant change in organizations, and that may mean high levels of resistance and skepticism. Any major training technology effort should involve a cost/benefit analysis to ensure that the expected outcomes justify the expenses and that the rationale is clearly made to all parties concerned.

There are many different cost/benefit models for HRD (ASTD; Spencer; Swanson and Gradous), but they all boil down to a comparison of what it will cost to deliver the training vs. the expected payoffs of that training. In the context of learning technologies, the cost components involve the initial and ongoing expenditures for hardware and software, as well as design and development of course materials. There are also labor costs associated with the delivery and implementation of courses: instructors, administrative and technical staff, consultants, etc.

The two categories of costs that should be less for technology-based training, and hence part of the benefits, are facilities and travel. Since technology should reduce the overall length of training by 25 to 50% relative to classroom instruction, trainees should require less time at training centers and lower hotel/per diem expenses. To the extent that technology makes distance-training possible, some facility and travel costs may be eliminated entirely. For example, suppose that a 10-day training program can be reduced to 6 days (40% reduction) through the use of technology. If that program involves 1000 employees annually and the value of each day's hotel/per diem is $75, the total cost savings of the reduced training time is 1000 persons × 4 days × $75/day = $300,000 for a year. Alternatively, imagine that the entire training program can be put in some form of multimedia or EPSS program and provided at the job site, eliminating the need for travel altogether. This eliminates $750,000 worth of hotel/per diem costs, plus the travel expenses, which are likely to be at least $250 per person ($250,000 total). In this case, the distance-training approach saves at least $1 million for a single 10-day training activity. These numbers do not even include additional savings due to training facilities not needed or reduced instructor/staff time.

Of course, money has to be spent on the design, development, and implementation of a program in order to displace the facility and travel costs. A multimedia version of the 10-day course to be used at a training center might cost $100,000 to design and develop and require another $100,000 worth of hardware and software. So the savings for this approach would be $300,000 – $200,000 = $100,000 annually. Development, hardware/software, and support costs for a distance-training version might add another $200,000, for a total cost of $500,000, resulting in savings of $1M − $500,000 = $500,000 annually. A real cost model would be much more detailed, with many cost categories and appropriate documentation for all costs based on historical expense data or vendor proposals.

The above costing example simply covers reductions in expenses, not the added value of benefits. To illustrate, consider that the value of an employee on the job in this example is $200 per day on average in terms of sales or services rendered. The value of 4 days on the job for 1000 employees instead of being away at training is 4 × 1000 × $200 = $800,000 in the first approach and 10 × 1000 × $200 = $2 million for the case of a distance-training program. These amounts are in addition to the cost savings already discussed. It is also possible to go farther with benefit calculations if we believe or have evidence that the outcomes of the technology-based training will actually result in better job performance. For example, suppose studies show that after the training, employees are able to perform twice as well as before, i.e., their average revenue per person is raised to $400 per day. Now the added value of the training is $200 per day for every working day following completion of the training. If there are 200 annual working days for each of the 1000 employees, the added value of the training is 200 × 1000 × $200 = $40 million. If the assumption about the improved performance is correct, this training will result in an extra $40 million of increased revenue per year to the corporation. While this kind of improved performance benefit requires a very effective technology implementation, it is possible and provides an example of the impressive financial outcomes that can be achieved under some circumstances.

Picking the Best Technology

Given that there are so many different technology choices, a major issue for every organization is what technology to select for a given learning/training application. Much analytical work has been conducted on the subject of media selection (Reynolds and Anderson; Romiszowski), and many organizations have developed sophisticated models for picking the best media for a given context. Indeed there are many variables to be taken into account, including:

1. Audience characteristics (background, age, total number)
2. Nature of the learning (type of skills/knowledge to be acquired)
3. Nature of the course (goals, extent of usage, timeframe to develop)

4. Availability of existing technology
5. Prior/current experience with specific technology
6. Availability of course content (already exists, new)
7. Management/staff/trainee support for technology use
8. Nature of the learning environment (classroom, learning center, workplace)
9. Availability of developers (in-house, consultants/vendors)
10. Availability of technology support/expertise
11. How will course be maintained/updated

These factors span instructional, logistical, and management considerations, all of which play a role in media selection decisions. The particular media selected must make sense instructionally in terms of addressing the learning needs of the employee. If the learning involves knowledge of facts about products, almost any approach will work; but if the trainee needs to learn procedural skills, some form of EPSS or simulation is desirable. Similarly, if discussions among participants and experts are needed, then some type of conferencing technology (audio, video, or computer) should be used. If there is a requirement to show examples of how equipment works or to show people interacting so behaviors can be modeled, television/video is a natural choice. Each medium has its strengths in terms of the kind of learning it facilitates (although there is a lot of overlap).

From a logistics perspective, different media involve different amounts of time, money, and expertise to utilize. For example, audioconferencing is a very inexpensive form of conferencing that can be organized with almost no preparation time and requires no special expertise on the part of trainers or trainees. On the other hand, satellite teleconferences require a great deal of preparation time (months), a lot of expertise to prepare and deliver, and are relatively expensive. Similarly, the development of interactive multimedia programs requires specialized design and development expertise and takes considerable time to create. In contrast, electronic documents (including those distributed via the networks) are fairly easy and quick to prepare, thanks to desktop publishing software and other authoring tools.

Finally, management considerations affect media selection. If some constituency opposes the use of a specific technology in training (or all technology), this factor has to be taken into account. If a particular technology has a poor history in an organization (rightly deserved or otherwise), it may be wise to choose another. On the other hand, if a certain technology has a good track record, or is already in use in the organization, this is a good reason to

Table 16.1   Strengths and Weaknesses of Different Technologies

	Strengths	*Weaknesses*
Print/	Inexpensive	Passive
Electronic publishing	Reliable	Updating required
	Worldwide distribution	
Television/Video	Dynamic	Development time/costs
	Large audiences	
Teleconferencing	Audience participation	Complexity (some types)
	Immediacy (real-time)	
Multimedia	Interactivity	Development time/costs
	Highly appealing	
Networks	Interactivity	Expertise required
	Updating required	
EPSS	High job relevance	Development time/costs
	Uses existing systems	Expertise required
Simulations	Authentic learning	Development time/costs
	Higher skill levels	Expertise required

consider it for other applications, even it is not the optimal selection from an instructional or logistics viewpoint. Some technologies are less threatening and more easily accepted than others in a given organization's culture, and this is an important factor in the decision.

Strengths and Weaknesses of Different Technologies

Table 16.1 summarizes the strengths and weaknesses of the major categories of technology discussed in this book. Clearly, taking into account all of the factors mentioned can make the decision process quite complicated. In practice, decision-makers tend to prioritize some factors over others and use these to make selections. Often cost considerations are an overriding concern — forcing the use of less expensive options. And, to tell the truth, media selections are frequently made on a very nonrational basis, e.g., a persuasive vendor presentation or simply copying what another organization is doing. The good news is that most types of training can be conducted with most types of technology, if the courses are designed properly and the material is developed well. Failing this, inappropriate choice of media with poor design and development can result in complete disasters — which are usually blamed upon the technology!

One other thing to keep in mind is that no single technology can do everything. For this reason, well-designed training programs normally involve a number of different media, each of which is used for different purposes. This also reduces the pressure to select the "perfect" media and the consequences of a single technology that doesn't work out. Of course, this makes the logistics of media preparation more complicated and raises the issue of how to integrate/coordinate different technologies in a single training program. If the program is well designed by experienced developers, this should not be a significant problem.

Why Don't We Weigh Them?

Guest author — Gloria J. Gery

Throughout the years, folks have developed measures for training effectiveness, satisfaction, and learning. All kinds of approaches from smile sheets to yardsticks have evolved. When I was running a data processing training organization at a large insurance company, I once got disgusted with the statistics I had to submit each month. As the functional manager, I was responsible for use of facilities, instructor resources, equipment, and assuring "value" for the training dollars spent on IT technical professionals and management — and the user community. My monthly report had to list items such as:

- Number of student days
- Student/instructor ratio
- Number of "no shows," drop-outs, and last-minute cancellations
- Dollars charged back to departments using training
- Percent utilization of facilities
- Cost per student-day
- Average "satisfaction" scores on our "smile sheet" evaluations
- On-time completion and on-cost development of new courses
- Actual versus planned operational budget expenditures.

At a meeting one day, I suggested a new measurement criterion. "Why don't we weigh the students and report on a cost per pound?"

A deep quiet came over the meeting. It was finally broken by a softly spoken question.

"What?"

I guess I was being given a chance to reconsider, but I didn't take it. "Why don't we install a scale in the entry way," I said, "like the one they use for cattle. We can have each student stand on the scale before entering class each day. We can then calculate the return on our investment by volume."

Needless to say, this attitude was a subject for much discussion both on that day and on my annual appraisal. While I wasn't exactly serious, the idea didn't seem any more irrelevant than some of the success indicators I was reporting on monthly. None of the measurements I was supposed to take asked if anyone learned anything or if our interventions changed their performance.

One of the men who worked with me was angry about my attitude. He said: "Do you know what your problem is?" (It's always a bad sign when somebody starts talking about 'what your problem is.')

"No," I responded.

"You're trying to get the right numbers instead of making the numbers come out right!" he said.

I am still working on a response to that one. But I long ago gave up trying to make the numbers come out right in favor of finding the best way to measure what we're trying to accomplish. Today, I encourage different measures. It's much easier to actually employ these assessments in a performance support environment because the connections between performance support in the actual work context is so much more direct than the distance between training events and work performance. That very statement says a lot, doesn't it?

Let me share some of the objectives and measurements that rule my work today:

- Decreased time to understanding
- Decreased time to performance
- Reduced performance cycle times (associated with a task, process, customer interaction, deliverable, creation, etc.)
- Reduced implementation costs (for a system, product, new process, etc.)
- Reduced support costs (number of coaches per group)
- Reduced hand-offs of work, calls, problems to others
- Increased customer satisfaction with organization representatives as measured by surveys, follow-up calls, complaint activity
- Quality improvements
- Ability to shift work to less experienced employees or to customers
- Reduced transaction costs
- Decreasing the gap between less experienced and star performers
- Competitive differentiation as reported by customers
- Organizational flexibility
- Increased performer confidence — and the confidence of those they work or interact with

When an organization can accomplish something like institution-alizing best practice into the work situation and make performance less a focus of individual competence and more a function of the environment itself, weighing people just doesn't come to mind for me. Does it for you?

Gloria J. Gery, a consultant in the fields of business learning and electronic performance support, is the author of *Electronic Performance Support Systems* (Gery Performance Press, 1991), the seminal book on EPSS. She can be reached at: gloria_gery@msn.com> This article originally appeared in *CBT Solutions Magazine*, May/June 1997. It is used here with permission of the author and the publisher.

Assessing Quality

In the past decade or two, there has been considerable attention paid to the issue of quality in organizations — largely due to the influence of W. Edwards Deming and the Total Quality Management (TQM) movement. The emergence of the Malcolm Baldridge National Quality Award and ISO 9000 standards further defined the criteria for quality in the U.S. and worldwide. By the beginning of the 1990s, most organizations had established some form of internal quality assessment process in which HRD usually played a critical role (Kaufman and Zahn; Lindsay and Petrick). And while quality assessment is not the same as performance improvement, there is a lot of overlap between the two approaches, including use of the same methodologies (such as needs analysis).

Figure 16.1 outlines some of the basic steps or processes common to most quality assessment approaches. At the beginning, there must be broad com-mitment at all levels of the organization (but especially top management) toward putting quality assessment procedures into place and following through on the changes they entail. This commitment also needs to empha-size customer satisfaction as the most important goal of the entire quality improvement process. Unless a high level of commitment can be achieved across the organization, quality assessment methods will not work.

Once commitment and consensus about quality assessment are achieved, the next step is to conduct needs analyses at the micro (individual groups or departments), macro (the entire organization), and mega (the marketplace and other organizations) levels. The purpose of these analyses is to identify customer

Figure 16.1   Model of the Quality Assessment Process

needs and gaps between present performance and desired outcomes. In some cases, a needs analysis will precede a full-fledged systems analysis which maps out all the functions and components of an organization or group in order to determine how things work (and fail to work).

The needs analyses lead to the development of quality specifications which identify outcomes that are to be achieved and measured in the pursuit of customer satisfaction. There will be many levels of specifications ranging from very general to quite specific, but all must be measurable in some fashion to be useful. Sometimes, specifications are aided by benchmarking activities in which the best efforts of other organizations are studied to identify reasonable goals or outcomes to aim for.

After quality specifications have been laid out, a variety of planning activities is needed to list the objectives to be achieved by various groups or programs and the responsibilities, resources, and timelines for doing so. This planning process needs to involve all members of an organization and customers to ensure that it produces the kind of changes needed to achieve the goals. Getting all "stakeholders" involved in the planning process is one of the most fundamental factors in the success of quality control efforts.

The last step of the process is the ongoing measurement of quality indicators to determine if quality improvement is being achieved. This involves much data collection, either through questionnaires, group discussions, work observation, or automatic measurements made by systems on product or service outputs. Customer satisfaction has to be assessed as well as job performance. Since most people seem to have an innate reluctance to be evaluated, this is one of the most difficult aspects of implementing a quality assessment approach.

The use of technology for training does not significantly change the nature of the quality assessment process — although it can make it much easier because it can be used to automatically collect performance measures and do data analysis. When it comes to quality assessment, technology is really not a critical aspect, although poorly implemented technology can be the subject of performance improvements!

Training Quality at Florida Power & Light

One of the most extensive (and successful) efforts to apply TQM principles organization-wide took place at Florida Power & Light (FPL) beginning in 1981 — and resulted in winning the Deming Prize in 1989. FPL employed basic statistical quality control techniques in all aspects of its operations and relied on three components: quality inspection teams, quality-in-daily work (QIDW) analysis, and policy deployment. Training activities, like all other business functions at FPL had to address these three components.

J.P. Magennis describes how this worked in the context of the Nuclear Power division of FPL. Training personnel worked on cross-functional teams to analyze specific task domains or problem areas and carry out the QI process. They also participated in the creation of QIDW control systems for each job process. Preparation of these control systems helped determine responsibilities for each job function and enhanced communication between training staff and plant customer groups. A variety of different data analysis techniques were used as part of the QI/QIDW processes including flow charts, bar charts, pie charts, pareto diagrams, histograms, and cause-and-effect diagrams.

Training staff were also part of the policy deployment process for the Nuclear Power division. The plant management teams identified specific areas that needed improvement related to overall goals and customer priorities; training staff focused their efforts on these areas as well. For example, in 1989, management identified accident prevention as a top priority, which was translated into improved simulator training for operators. By paying particular attention to quality improvement of this aspect of training, the quality goals of each plant and the entire Nuclear Power division were supported.

In addition to the various data analysis techniques practiced continuously for the QI/QIDW processes, a great deal of interaction is needed among all levels of training and management to make TQM work. For example, there are regularly scheduled meetings between corporate and site training managers, between plant management and training staff, and between line supervision and students. Other line management responsibilities include observation of training in progress, periodic meetings with students, monitoring attendance and exam performance, and reviewing training plans, schedules, or changes. Having operations staff involved with training and training staff involved in operations decisions is a critical foundation for quality improvement in learning activities.

Magennis, J. P. (May/June 1995) Training quality: before and after winning the Deming Prize. *Educational Technology.*

Evaluation Matters

There are many different ways of evaluating training and technology, ranging from qualitative case studies (Bowsher) to various formal methods and techniques (Baker and O'Neil; Flagg). In this chapter, we have looked at some of the most important aspects in the organizational setting, namely effectiveness, cost/benefit, and quality assessment. To use technology successfully in the context of performance improvement, all of these major forms of evaluation need to be practiced. However, it matters less which specific techniques or methods are used so long as some evaluation activities are conducted for all learning processes and the results are analyzed routinely.

There has been a long-standing tradition in the training world to rely on student satisfaction surveys (often called "smile sheets") and knowledge gained as the primary measure of effectiveness. While evaluation of student satisfaction and knowledge improvement is helpful, it is only one measure, and not one that provides much useful information relative to performance improvement. To make good decisions about the use of training technology and to determine the learning outcomes associated with them, HRD professionals need to adopt the more sophisticated measurement processes that include measuring improvement in job performance and determining if there has been significant improvement in the organization, e.g., profits, customer satisfactions, etc. (Kirkpatrick).

Finally, selection of learning technology has often been based on fads or vendor persuasion. This is a good way to waste money that no successful organization can afford. Determining the appropriate technology for a given training application needs to involve a cost/benefit process that identifies the financial and HRD payoffs as well as the investments required for the design, development, and ongoing support of that technology. There also needs to be a careful and accurate cost estimation methodology to identify the complete costs of implementing a given technology.

Summary of Key Ideas about Technology Evaluation

- Research shows that technology-based training can be as effective as classroom instruction and usually results in time savings.
- Many forms of technology, such as simulation, have been shown to be more effective than conventional forms of training.
- Learning via technology and classroom instruction probably teach different kinds of things even when they involve the same content.
- Analyzing the cost/benefits of training technology is important because significant financial investments and organizational change are usually involved.
- The two cost categories that technology usually reduces are travel and facilities, whereas equipment costs increase.
- The greatest value of technology is often in terms of increased job performance (i.e., lost opportunity costs).
- In selecting the best technology for a given application, a number of instructional, logistical, and management considerations need to be taken into account.
- While different technologies have their strengths and weaknesses for certain kinds of applications, design and implementation of the training are typically more important than the choice of technology.
- Use of technology needs to be closely related to quality assessment in an organization and include processes such as needs analysis, quality specifications, and measurement procedures.

■ Evaluation is particularly important in the context of technology use because it is highly susceptible to fads and marketplace trends.

References

ASTD. (1997) *How to Conduct a Cost-Benefits Analysis.* Info-Line Publication 90-007. Alexandria, VA: ASTD.

Baker, E. and O'Neil, H. (1994) *Technology Assessment in Education and Training.* Hillsdale, NJ: Erlbaum.

Bowsher, J. (1988) *Educating America: Lessons learned in the Nation's Corporations.* New York: John Wiley & Sons.

Flagg, B. (1990) *Formative Evaluation for Educational Technologies.* Hillsdale, NJ: Lawrence Erlbaum.

Gery, Gloria J. (May/June 1997) Why don't we weigh them? *CBT Solutions Magazine.*

Kaufman, R. and Zahn, D. (1993) *Quality Management Plus: The Continuous Improvement of Education.* Thousand Oaks, CA: Corwin Press.

Kirkpatrick, D. (January, 1996) Techniques for evaluating training programs. *Training and Development.*

Lindsay, W. and Petrick, J. (1996) *Total Quality and Organizational Development.* Boca Raton, FL: St. Lucie Press.

Magennis, J. P. (May/June 1995) Training quality: before and after winning the Deming Prize. *Educational Technology.*

Reynolds, A. and Anderson, R. (1991) *Selecting and Developing Media for Instruction* (3rd ed.). New York: Van Nostrand Reinhold.

Romiszowski, A. (1988) *Selection and Use of Instructional Media* (2nd ed.). New York: Nichols.

Spencer, L. (1986) *Calculating Human Resource Costs and Benefits.* New York: John Wiley & Sons.

Swanson, R. A. and Gradous, D.B. (1988) *Forecasting the Financial Benefits of HRD.* San Francisco: Jossey-Bass.

17 Impact of Culture and Globalization on Technology and Learning

In today's rapidly globalizing marketplace, understanding and valuing the nuances and synergy of cultural differences is critical for both employee satisfaction and corporate success. Jack Welsh, CEO of General Electric, recently commented that companies "must either globalize or they die." David Whitwam of Whirlpool echoed these sentiments when he emphasized the importance of creating organizations "where people are adept at exchanging ideas, processes, and systems across borders."

Managers and HRD professionals who have worked internationally have long recognized the fact that people live and learn differently because of culture; they are managed and motivated differently because of culture; they think and react differently because of culture. Why? Because our cultures have taught each of us a specific "correct" way to think, act, and do things which is different from every other culture's way of thinking, acting, and doing things.

And what does the added aspect of technology bring to the cross-cultural equation? Does technology also have to be used differently because of culture? Does globalization affect how technology is implemented across distances, times, and borders? Or will globalization blur national boundaries and ethnic communication systems and lessen the importance of culture?

These are the key cultural, global, and technological questions we will explore in this chapter. First, however, we need to define and understand several basic concepts relative to culture and globalization.

Culture — Different Ways of Doing Things

Culture can be described in many ways. Most definitions, however, contain three essential elements: 1) it is a way of life shared by all or almost all members of a group 2) that older members of the group pass on to younger members, 3) which shape one's perceptions and behaviors. Culture provides systematic guidelines for how people should conduct their thinking, doing, and living. Thinking (ideas) encompass values, beliefs, myths, and folklore. Doing (norms) include laws, statutes, customs, regulations, ceremonies, fashions, and etiquette. Living (materials) refers to the way one interacts with machines, tools, food, natural resources, and clothing (Bierstadt; Pace, Smith and Mills). There are ethnic cultures, national cultures, corporate cultures, and an emerging global culture.

Hofstede distinguishes a group's culture both from the universality of human nature and from the individuality of each person (Figure 17.1). *Human nature* is what all human beings have in common and is inherited with one's genes. The human ability to feel anger, fear, love, joy, sadness, loneliness, to think, to learn, to work and play, are all part of this human programming. However, what one does with these feelings, how one expresses self or behaves is modified by the *culture*, is taught to us as we enter that environment. *Personality* is how an individual uniquely acts and is modified both by the influence of culture and one's personal experiences.

An illustration of the three levels would be to look at the functions of eating or learning: a) all *humans* eat or learn in their daily lives; b) however, each *cultural* group eats or learns in different ways; c) finally, *individuals* in a cultural group may eat or learn similarly as well as differently from other members of their cultural group.

It is also important to remember that culture is multilayered (Figure 17.2). *Practices,* which include behaviors, symbols, rituals, and artifacts, are more visible to someone outside the culture. They are more easily influenced and changed than the core of culture which is formed by the *values* and underlying *basic assumptions* that are not so easily recognized or understood by outsiders. These values are among the first things children learn — not consciously, but implicitly.

Simply put, culture provides people with a meaningful context in which to meet, to think about themselves, and to face the outer world (Trompenaars; Marquardt). It is important to realize that culture is logical and rational to the members of the culture, but often appears to be irrational or illogical to someone outside the culture. This represents the greatest challenge to anyone trying to work or live, or to learn or apply technology, in another culture (Storti).

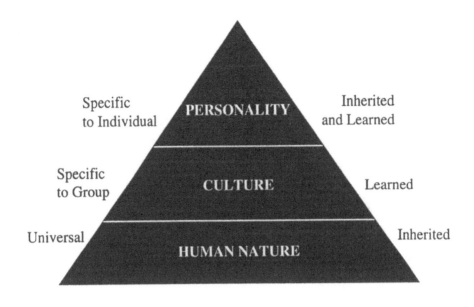

Figure 17.1   Culture as Distinguished from Human Nature and Personality

Figure 17.2   Layers of Culture

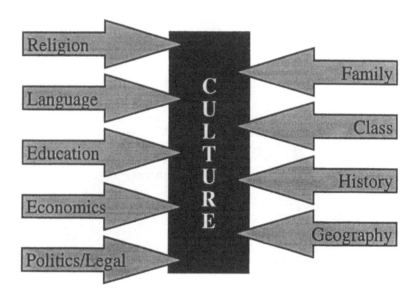

Figure 17.3   Factors of Culture

Factors of Culture

There are nine factors that create a culture and, in turn, are influenced by the culture: 1) religion, 2) language, 3) education, 4) economics, 5) politics/law, 6) family, 7) class structure, 8) history, and 9) geography/natural resources. What distinguishes one culture from another is not the presence or absence of these factors, but rather the patterns and practices found within and between these factors (Marquardt and Engel). Each of these factors can and does influence how a cultural group responds and reacts to action learning (Figure 17.3).

Western and Non-Western Cultural Contrasts

Robert Kohls, a noted American cultural anthropologist, contrasts the values differences between Western and non-Western cultures in the following manner:

Western Cultural Values	Non-Western Cultural Values
Individualism	Collectivism/group
Achievement	Modesty
Equality/egalitarianism	Hierarchy
Winning	Collaboration/harmony
Guilt (internal self-control)	Shame (external control)
Pride	Saving face
Respect for results	Respect for status/ascription
Respect for competence	Respect for elders
Time is money	Time is life
Action/doing	Being/acceptance
Systematic/mechanistic	Humanistic
Tasks	Relationships/loyalty
Informal	Formal
Directness/assertiveness	Indirectness
Future/change	Past/tradition
Control	Fate
Specific/linear	Holistic
Verbal	Non-verbal

These value differences provide a mirror as well as a window that can explain why people of Western cultures will tend to react differently from people of non-Western cultures to technology, learning, management, and to the workplace. Here are a few examples of the impact of such contrasts in learning situations.

1. In non-Western cultures a high value is placed on visible status. The quality of the instruction, the mode of technology used, and the status of educational resources will determine the importance with which the instruction is perceived and who will attend.
2. The social expectations of many cultures require more interaction with the instructor, hopefully face-to-face so as to meet the affiliative needs of the learner.
3. Some cultural groups allow little or no intermingling of workers or learners of different ages, genders, or rank in training programs.

Globalization and Culture

Over the past 100 years, companies have undergone a corporate evolution from domestic markets toward globalization. Domestic companies first entered the *international* phase when they began importing and/or exporting parts or products. As volume of sales or manufacturing increased to substantial levels, the companies established manufacturing and sales centers in individual countries or regions — the *multinational* phase. Only in the past 20 years have companies been able to become *global*; i.e., operate as if the entire world were borderless and a single entity. Global companies, thanks to technology, can be centralized and decentralized (global reach and local touch). They are integrated so that they can link and leverage their strengths across borders. Globalization has occurred when the organization has developed a global corporate culture, strategy, structure, and communication process (Rhinesmith; Marquardt and Reynolds).

One of the reasons global companies are so powerful is their capacity to mobilize the synergies and strengths of various cultures. This capability leads to multiple perspectives, the development of a wider range of options and approaches, the heightening of creativity and problem-solving skills, and increased flexibility in addressing culturally distinct clients and partners. Thinking and operating globally will be critical to organizational survival and growth in the 21st century (Casse; Gannon; Marquardt and Snyder; Herriot and Pemberton).

Globalization does not erase culture. As a matter of fact, research has shown that employees of global companies retain their cultural beliefs and practices more strongly than employees of domestic companies (Abdullah). Although there are some global cultural values which are emerging on a worldwide scale (e.g., service orientation, value of teamwork, continuous learning, respect for individuals), cultural values are not disappearing — even though people around the world may wear Levi's, eat pizza, listen to Western music, and watch CNN. As noted above, these only represent behavioral changes, not core cultural values and basic assumptions.

Developing Learning Programs for Multicultural Settings

In Chapter 5 we described how technology affected every aspect of training design and delivery. Likewise, as we will note here, culture and globalization also affect every aspect of the learning program. The interplay between technology and the learning program can be significantly affected by the cultural setting. Let us examine some specific ways in which culture affects the design, development, and delivery of a learning program as well as possible steps or precautions the HRD professional might consider.

Designing the Curriculum

In designing the curriculum, one should be aware of the learning traditions of the local culture(s) because they impact the learning styles of the participants. The participants' learning style will affect the structure and sequence developed in the lesson plan and curriculum as well as the delivery methodologies.

Although learning styles are influenced by a variety of cultural factors and differ from country to country, there are some fundamental differences between the learning style of Americans (and northern Europeans and Australians) and that of the rest of the world. Americans have a cultural system, learning style, and reasoning preference that tends toward the inductive task or problem-centered approach. Most of the rest of the world, however, has a strong preference for deductive, topic-centered reasoning. These deductive cultures do not tend to apply factual data in the multiple ways that people in the inductive cultures do every day of their lives.

In preparing a curriculum for deductive cultures in which trainees are not accustomed to moving from specifics to generalizations, the deductive approach and topic-centered structure should be initiated. Perhaps, after a period of time, an inductive, experiential approach can be introduced.

Another aspect of curriculum development is the scheduling of time allotted to various content areas and activities. Some useful suggestions:

- Do not schedule too tightly because learners in most cultures want considerably more time for discussion and exploring.
- Take into account the values, styles, and attitudes of the learners; do not limit team activities to 15 minutes when the cultural perception of agreement is harmony and consensus, and not majority rule.
- Bilingual programs need much more time than those that are monolingual.
- If English is the second language, schedule more frequent breaks to allow rest for the trainees, who are attempting to comprehend and speak in English.
- Allow time for socializing and building relationships.

Many global trainers, in comparing curriculum designs among French, Germans, and Americans, find that Germans feel quite comfortable with training developed in the U.S. because American curricula are highly structured, have a tight logical flow, and employ the rapid pace enjoyed by Germans. The French, on the other hand, prefer a slower pace and find American training lessons too cursory. The French enjoy discussing and arguing the merits of a subject. A one-day program in the U.S. would generally need to be designed for two days in France.

Selecting Training Methodologies

There are literally more than 100 different training methodologies that one could choose in delivering a learning program, ranging from the didactic, trainer-directed approach to the experimental, learner-centered, and skill-focused approach. American trainers have tended to emphasize experimental learning, adhering to the adult learning theories of Knowles, Bradford, and others.

For workers in most cultures, however, these self-directed learning methodologies can be very uncomfortable and run counter to cultural mores. Participants from Asian and Arab cultures feel much more at ease learning by rote and prefer to observe the instructor demonstrating a skill rather than face the possibility of being seen as foolish through risk-taking and learn-by-doing methodologies.

Cultures that a) expect their instructors to teach with authority and power, and b) to be experts with absolute truths will obvious prefer training methods that are more teacher-centered. Cultures that are more egalitarian, participative, and seek less structure will prefer training methods that are more learner-centered.

Distance learning is both enjoyed and feared by some cultures. Although technology may be less likely to force them to publicly interact, it lacks the affiliative flavor of a live instructor.

Depending on the culture, specific methodologies may be effective or ineffective. In Asian countries, for example, experienced global trainers discourage the use of role playing and structure experiences because these cultures find it difficult to place themselves in the shoes of others or to confront each other because of the high cultural value placed on conflict avoidance and authority relationships. In addition, risk-taking through role playing in countries like Japan and Korea is taboo, because no one wants to be a fool or stand out (Miller). The Japanese are accustomed to lectures, note-taking, and respectfully asking questions of teachers. Learners attempt to soak up the information like sponges and feed it back verbatim.

Games can also be ineffective because Asians separate game playing from the serious business of learning. In Asia, the best participative training methodologies are small group discussions and case studies carried out between participants of similar age and status. These should then be followed by a report of results gained by each small group to the total group by representatives of each group. Why? Because this allows the representatives to discuss the process freely without any individual member having to take personal responsibility or accolades for them.

Structured experiences such as "Star Power" or role playing that reverse the normal power positions can become very uncomfortable for participants from many cultures. Asians attach great value to their positions and status differences. Exercises that strip them of these trappings tend to cause embarrassment, confusion, and loss of face for all participants, and thus reduce learning.

The use of interactive video as a training strategy may not be appreciated in countries like Japan (where it was invented) and other collectivist societies because it is a method that focuses on the individual. The Japanese prefer to learn and work in teams.

The use of debate or extensive verbal methodologies will be less effective in Chinese and Japanese cultures because they conflict with at least three cultural beliefs by:

1. Placing human-centered hierarchy over propositional truth
2. Over-emphasizing definition and distinction, whereas the Chinese and Japanese languages create a world view more given to imagery and ambiguity
3. Valuing oral communication over meditation, reflection, and thought

In a similar vein, the use of case studies and analysis in Arab countries may be ineffective because Arab culture encourages verbal comments by only the leader or manager of the group and not the individual participants. In Saudi Arabia and other Gulf countries, paper exercises and games are thought to be activities for school children; the preferred methodology for learning is group discussions.

Germans tend to prefer methodologies that are orderly, systematic, detailed, and analytical (case analysis), while the French seek lively, witty activities (e.g., brainstorming).

Since the eventual utilization of the experimental methodologies identified above is so critical to developing certain skills and competencies, it is best to perhaps begin more formally, through lecturing and modeling. Only after gaining the confidence and comfort of the trainees, should the instructor begin employing more learner-centered methodologies.

Developing Learning Materials

During this step of the training design, the necessary learning materials are written, adapted, and/or produced. These may include workbooks, handouts, intranet materials, instructor guides, audio and videotapes, and computer software. The process of creating new materials or adapting already developed materials for export to other cultures is not a simple task. The HRD professional must "not only translate materials, but also be sure to transpose all that we have learned to a totally new environment in a way that gains acceptance in the local culture" (Morical and Tsai).

Materials developers should assume cultural differences rather than similarities. They should treat each cultural environment as different and requiring culturally appropriate learning materials that will not be seen as offensive or be resisted by the learners.

The following steps are recommended in creating training materials to use in a another culture:

1. Local instructors and a translator, ideally someone who is bicultural, observe a pilot program and/or examine written training materials.
2. The educational designer then debriefs the observation with the translator, curriculum writer, and local instructors.
3. Together they examine the structure and sequence, ice breaker, and materials.
4. They identify stories, metaphors, experiences, and examples in this culture that might fit the new training program.
5. The educational designer and curriculum writer make changes in the training materials.
6. The local instructors are trained to use the materials.

7. Materials are printed only after the designer, translator, and native-language trainers are satisfied.
8. The language and content of the training materials are tested with a pilot group

Motorola is perhaps the best global corporation in acculturizing training materials so that knowledge and skills can be transferred promptly and effectively in the local culture. Subject matter experts, technical writers, customers, course developers, cultural anthropologists, translators, instructors, applications consultants, and program managers are all involved in developing training materials for non-American learners.

Adapting existing training materials

In adapting training, one should avoid using training materials that represent American culture or imply that the American way is the best or only way. If too American-focused, these materials may not tell the trainees how to apply the knowledge and skills in their own environments.

Illustrating concepts with examples that are familiar and acceptable to the trainees makes learning much more productive and enjoyable. Exercises should be tailored to the local culture, for example, through the use of local names, titles, and situations whenever possible. One should modify the situations, change case studies, and rewrite applications when taking training materials to another culture.

The degree of adaptation required depends on what is being taught. The more technical the topic, the less need for change. For example, if you want to teach someone to operate a particular machine, there will be need for very little adaptation. On other topics, however, such as communication and supervisory skills, the amount of training material to be adapted will be considerable (Odenwald).

Translation

Translation is more than translating word for word the content from one language to another. To the extent possible, the values and local applicability must also be translated. Translating instructional materials is a difficult process, especially when it involves formulating new words to conform to both good language usage, cultural meaning, and to the learning objectives of the training program.

It is important to remember that there are words that are not directly translatable to another language (e.g., "no" into Thai, the many "you's" of Thai into English). There is also not an automatic interchangeability of technical terms (e.g., there are no Arabic words for "metal tubing" or "jet propulsion").

It is important, therefore, to study the language — vocabulary and terminology. Gaining agreement with the local people on terminology and a glossary will help in preparing training manuals and workbooks. An excellent source for terminology is the International Labor Organization, which has published a series of glossaries for technical and vocational training in a number of different languages.

A final aspect of translation is the level of language skills, written and oral, of the learners. Training materials should be prepared according to their language capabilities.

Specific cultural guidelines when developing training materials

There are also a number of culturally-specific guidelines to consider as you develop training materials.

- Avoid culturally inappropriate pictures or scenarios, e.g., women for Arab cultures.
- Use plenty of graphics, visuals, and demonstrations if the trainees are learning in a second language.
- Provide many handouts and instructional materials; these are highly valued and even displayed in many cultures.
- Beware of corporate ethnocentrism in which the company must be presented exactly the same all over the world.

In many cultures, especially Asia where a high value is placed on clear and specific instructions, training materials must be well organized and unambiguous. The materials should include written examples for all worksheets required to be completed by participants, explicit written instructions for any exercise, and complete and accurate summaries of all lectures. Without these specific instructions, participants may become agitated by the lack of direction and, often covertly, blame the trainer for any negative outcomes resulting from the training experience.

Acculturization of Technology-Based Learning Materials

Courseware

Technology-based courseware needs to be especially "acculturized" for it to be effective outside the culture in which it was developed. Nielson emphasizes the cross-cultural aspects of user interface design, noting that a number of global software companies (i.e., Microsoft, Apple Computer) have developed clear design guidelines for courseware programmers to follow.

Unlike the live instructor who is using an oral medium which can be acculturized "on the run" by an instructor, learning via technology depends on the written and graphic communications between a program and a learner. According to Reynolds, there are a number of areas to consider in adapting computer programs to other cultures.

Space

Some languages require more printed space than others to convey the same information. For example, German requires approximately 30% more space than English. Extra space must therefore be reserved simply to provide sufficient room for the new language.

Format

Problems can arise when the target language does not lend itself to an acceptable presentation within the available space format. For example, Arabic is written from left to right, so the graphic should go on the left of the text rather than on the right as in English.

Simplicity

Although creativity is valuable for making a program interesting and effective, it is important to keep a lesson reasonably simple so that it can be more easily adapted.

Clarity

An explanation or narrative that is obscure in English will be just as obscure in Indonesian or Swahili.

Standardization

Standards protect against practices that make adaptation difficult. Specify the size of the borders and locations where certain information will always appear on the learner's display.

Cultural Relevance

Use narratives and examples that fit the culture and environment. Baseball, for example, is not understood in most parts of the world.

Jargon

Jargon does not lend itself to translation, and employing current slang in training materials or presentation is usually neither necessary nor effective.

Acronyms

Acronyms create confusion. What is easily pronounced or understood in English may be unfathomable in another language.

Humor

Humor is always potentially dangerous because it can leave the learner wondering what the point is or, worse, insulted. It is also difficult to translate.

Videotape

The video content must fit the culture and be sensitive to local customs and behavior. What do you do if you are producing a video that you plan to utilize in two or more cultures? Do you dub in the local language, use subtitles, or reshoot all or part of the video? It depends on two factors: 1) the norm for the culture you are working in, and 2) the skills you are trying to teach.

Based on the cultural norm, for example in France, you would generally dub (the French are accustomed to watching American movies), while in

Scandinavian countries you would use English subtitles (that is the custom there).

If the topic you are developing is "interpersonal skills," you would reshoot the entire video because this skill is so very cultural. "Business etiquette" would also need to be reshot due to unique cultural differences. For example, Americans talk business during a coffee break, while the French are there to drink coffee.

Charles Tweedly of Otis Elevator describes a "wake-up call having to do with language and cultural issues." For example, the French dialect spoken in Quebec is not the same as that spoken in France (and practiced by most translators). The solution was to use local translators as opposed to U.S.-based speakers of a particular language.

Tweedly points out that the training software at Otis is designed with minimal text incorporated and with no spoken soundtrack. When a division in India, for example, receives the script, the local managers translate it. His advice: be sure to have hardware and software uniformity; limit the amount of text on screen so only the soundtrack requires local translation.

Culture and Globalization Shape Technology

In areas that require human interaction, culture will always have an impact. Activities taking place in another culture, in multicultures, or globally will overlap with strongly held national norms or values. Each culture has different ways of valuing and implementing work, learning, management, customer service, teamwork, etc. Different technologies by their nature may imply values such as cooperation, open exchanges, control, equity, individualism, hierarchy, etc. (Gerber, Callahan).

Not all aspects of technology, however, are affected by culture or globalization. For example, the use of a machine such as a Xerox 5095 copier/stapler would seem to be similar regardless of the country in which the machine is located. Many manufacturers of technology have already made adjustments to adapt to cultural and language differences. As an illustration, the latest pocket-size pagers are capable of receiving alphanumeric messages and include a disposable battery with an average lifetime of up to one month. To facilitate worldwide communications, message displays with international character sets are common.

The Most Important 10%

Akio Morita, former Chairman of Sony, recently commented that "culture impacts products, services, and operations by only 10%, but this is the most important 10%. This 10% determines success or failure." Like a growing number of global leaders, he recognized and understood the crucial importance of acculturizing technology, learning, and the workplace to the values and cultural practices of the local people and environment.

Goodman and Darr also acknowledge that global companies face the difficult task of getting employees to effectively "collaborate and share expertise unbounded by language, culture, space, and/or time." Yet unless we can achieve this synergism, we will not be very successful in the highly competitive global marketplace. Acculturized technology combined with acculturized materials and learning programs can elevate the competencies of multicultural groups in multicultural settings.

Summary of Key Ideas About Impact of Culture and Globalization on Technology

- All companies and workplaces are influenced by the forces of culture and globalization.
- Culture guides the way we think and act; it includes external behaviors and internal values and basic assumptions.
- Companies have evolved from domestic to international to multinational to global stages.
- Culture impacts the way people learn and prefer to be trained; therefore, learning programs need to be designed, developed, delivered, evaluated, and administered differently for different cultural settings.
- Methodologies effective in Western cultures may be totally ineffective in non-Western cultures.
- Technology needs to be adapted to accommodate space, format, clarity, jargon, and cultural relevance.
- Video may need to dubbed, subtitled, or re-shot depending on the culture.
- Different technologies, by their nature, may imply values such as cooperation, open communication, control, equity, hierarchy, individualism, etc.

References

Abdullah, A. (1996) *Going Glocal.* Kuala Lumpur: MIM Press.

Bierstadt, R. (1963) *The Social Order: An Introduction to Sociology.* New York: McGraw-Hill.

Callahan, Madelyn. (March, 1989) Preparing the new global manager. *Training and Development.*

Casse, P. (1982) *Training for the Multicultural Manager.* Yarmouth: Intercultural Press.

Gannon, Martin. (1994) *Understanding Global Cultures.* Thousand Oaks: Sage Publications.

Gerber, B. (September, 1989) A global approach to training. *Training.*

Goodman, P. and Darr, E. (1996) Computer-aided systems for organizational learning, in *Trends in Organizational Behavior,* Cooper C. and Rousseau, D. (Eds.), New York: John Wiley & Sons.

Herriot, P. and Pemberton, C. (1996) *Competitive Advantage Through Diversity.* Thousand Oaks, CA: Sage.

Hofstede, G. (1991) *Cultures and Organizations.* London: McGraw-Hill.

Kohls, L. R. (1981) *Developing Intercultural Awareness.* Washington, D.C.: SIETAR.

Marquardt, M. and Engel, D. (1993)*Global Human Resource Development.* Englewood Cliffs: Prentice-Hall.

Marquardt, M. and Reynolds, A. (1994) *The Global Learning Organization.* Burr Ridge: Irwin Professional Publishing.

Marquardt, M. and Snyder, N. (1997) How companies go global. *International Journal of Training and Development.*

Miller, V. (1993) *Guidelines for International Trainers in Business and Industry.* Boston: International HRD Press.

Morical, K. and Tsai, B. (April, 1992) Adapting training for other cultures, *Training and Development.*

Nielsen, J. (1991) *Designing User Interfaces for International Applications.* New York: Springer.

Odenwald, S. (1993) *Global Training: How to Design a Program for the Multinational Corporation.* Homewood: Irwin Professional Publishing.

Pace, W., Smith, P., and Mills, G. (1991) *Human Resource Development.* Englewood Cliffs, NJ: Prentice Hall.

Reynolds, A. (1984) Adapting courseware, in *Technology Transfer,* Reynolds, A. (Ed.) Boston: HRD Press.

Rhinesmith, S. (1992) *A Manager's Guide to Globalization.* Homewood: Business One Irwin.

Trompenaars, F. (1993) *Riding the Waves of Culture.* Oxford: Economist Press.

Tweedly, C. (May, 1995) Multimedia delivery makes unified global training a reality. *Training Directors Forum Newsletter.*

18 Best Practices in Learning Technology

I n Part II, we described a variety of learning technologies — teleconferencing, interactive multimedia, television and video, electronic publishing, and simulation and virtual reality — that enable an organization to enhance the speed and quality of learning in the workplace. A growing number of world-class organizations have begun employing these learning technologies in an effort to maximize human performance and build corporate capabilities for global success. In this chapter, we will highlight three companies that are leaders in utilizing technology for learning — Federal Express, Ford Motor Company, and Nortel.

Federal Express

Federal Express is the world's largest express transportation company, delivering over 3 million items in 200 countries each working day. Headquartered in Memphis, Tennessee, the numbers at FedEx are large and growing — over 140,000 employees, daily flights into more than 325 airports, 2000 staffed facilities, and more than 30,000 drop-off locations. The company prides itself in setting "the standards in the shipping industry for reliability, innovative technology, logistics management, and customer satisfaction." FedEx has received numerous awards, including the Malcolm Baldridge National Quality Award.

Under the guidance of CEO Fred Smith, Federal Express has made a conscious and deliberate effort to use technology in improving the speed and quality of learning within the organization. FedEx leaders are quick to point out that technology has significantly boosted the company's intellectual capacity, agility, and resourcefulness.

Huge Investments in Learning Technologies

Federal Express has made enormous investments in building its learning technology resources — more than $40 million in 1200 systems in 800 field locations. Each location is stocked with 30 interactive videodisc programs, which have been used to train many of FedEx's 35,000 couriers and customer service employees.

Focused employee training using technology has been going on at FedEx for more than ten years, but in 1995 the company launched a new interactive training system using multimedia workstations made by Silicon Graphics. These screens combine TV-quality video with text, graphics, and voice to teach basic interaction skills such as customer-contact methods and the features of its service categories for its couriers and customer-service agents. People are trained at a pace that's more personal for them and more customized than the stick-and-pointer classroom. A certain amount of training is required annually, but because it is done using interactive multimedia, there is more flexibility than in the past when classroom time had to be scheduled. Now training can occur at the beginning or end of a shift, or whenever the individual can best fit in what will be very personalized instruction.

In recent years, Federal Express replaced some of its classroom training programs with a computer-based training system that uses interactive video on workstation screens. This training system can capture and interpret input from learners to determine whether a task is being performed correctly. If a learner makes a mistake, the system recognizes the error, points it out, and shows the proper method.

The interactive video instruction system presents training programs that combine television quality, full-motion video, analog audio, digital audio, text, and graphics, using both laser disc and CD-ROM. Learners can interact with the system using a touch screen or keyboard.

The interactive video training closely correlates with job testing. Using the system, employees can study their job, company policies, and procedures, and brush up on customer service issues by reviewing various courses. Currently, there are over 1200 interactive video instruction units placed at more than 700 Federal Express locations. All workstations are linked to the Federal Express mainframe in Memphis. Each location has 21 video disks that make up the customer-contact curriculum. There is virtually no subject or job-related topic that the customer-contact workers cannot find on the interactive video instruction platform.

Once the CD-ROM courseware is written, FedEx knows that it is imperative to keep it updated. The work force relies on the fact that the system provides accurate and current information. For them, out-of-date information is worse than no information at all. For this reason, a new CD-ROM is sent to each location every six weeks. This CD-ROM updates the curriculum through text, PC graphics, and digital audio. Over 1000 updates are made on an annual basis.

Here is a current list of Federal Express' interactive laser disk and CD-ROM courseware:

Accepting International Priority Shipments
Agent as a Salesperson
Assisting Customers with Other Departments
Calculating Rates
Computer Resources
Dangerous Goods: Acceptance Made Easy
Defensive Driving
Delivering Packages
Domestic Documents
Domestic Services and Packaging
Effective Customer Communication
Fundamentals of Customer Service
International Products and Services
Package Tracking Inquiry System
Personal and Vehicle Safety
Picking Up Packages
Special Services

Performance Improvement Through Interactive Video Instruction

FedEx recently created a mandatory performance-improvement program for all of the company's employees who deal with customers either face-to-face or over the phone. The primary goals of this program were: 1) to completely centralize the development of training content while decentralizing delivery, and 2) to audit the employees' ability to retain what they learned.

The pay-for-performance program consists of job knowledge tests that are linked to an interactive video instruction (IVI) training curriculum accessed on workstations in more than 700 locations nationwide. More than 35,000 Federal Express customer-contact employees around the country are required to take the job knowledge tests annually via computer terminals at their work locations. The tests, which measure employees' knowledge in their specific jobs, correspond with employees' annual evaluations. In fact, the results of the tests make up approximately one tenth of the employees' performance ratings.

By testing customer-contact employees on product knowledge services, policies, and various aspects of their jobs, FedEx obtains two major benefits according to William Wilson, manager of training and testing technology, namely:

1. All employees operate from the same book, ensuring that all customers will receive accurate and consistent information during each transaction. This helps the company maintain its high service levels and commitment to quality.
2. Managers have an objective way to measure job knowledge for all customer-contact employees.

Federal Express provides many incentives for workers to increase their learning quickly. For example, employees are paid for two hours of test preparation prior to each test, two hours of test time, and two hours of post-test study time.

The current average amount of time that workers use the Interactive Video Instruction program is approximately 132,000 hours per year. Compared to traditional training, this equates to approximately 800 one-day classes with 20 employees per class. Yet no trainers are necessary, and no travel costs are incurred.

Quality Using Electronic Systems Training

Federal Express also developed a test program called QUEST (Quality Using Electronic Systems Training) to ensure that all of the learning tests are valid, relevant, fair, and meet appropriate learning standards. This was done by creating focus groups composed of trainers, managers, and job incumbents. The focus groups designed each of the tests, which consisted of multiple-choice questions pertaining to all important aspects of employees' jobs.

Based on the members' collective knowledge, the focus groups created surveys listing the critical tasks for each job. Workers within those jobs were then asked to rate the tasks in order of importance. Focus groups then wrote the test questions based on those issues, being careful to include only questions that directly pertained to activities in which the workers engage.

The final step before implementation of the technology-based learning was to conduct some pilot testing. At this phase, subject matter experts and an on-staff industrial psychologist examined any questions that might be construed as unfair based on the number of workers who missed them. The entire process — from focus-group formation through test validations and implementation took approximately 15 to 18 months.

In order to keep the tests timely, FedEx had the original focus groups meet quarterly to discuss existing test questions to ensure that they were still valid. The groups also spent time in writing new questions. Over a period of time, FedEx has built up a bank of several hundred questions for each test. If questions are eliminated, they are pulled from the bank, and equally weighted questions are inserted from the same topics.

Federal Express has found that the QUEST automated program saves hours in clerical and administrative activities because the computer does all the scoring, recordkeeping, item analysis, and score reporting. Additional features of the program are real-time registration, real-time test score reporting, and item analysis.

EPSS Enhances Speed and Quality of Learning

Federal Express customer-service representatives get thousands of telephone calls a day, each of which demands ready answers. In the past, FedEx representatives would have handed questions by passing customers and problems along to another representative. EPSS, however, has enabled FedEx to resolve problems "immediately and proactively without passing off any customers," says Bart Dahmer, manager of technology services and technical training at Federal Express in Memphis, Tennessee.

EPSS makes it possible for the customer service representatives to access one computer application without closing down another. For example, representatives will not have to exit "billing" to get into "customer service." They will be able to retrieve information from several databases and promptly put it onto their computer screens so they can address a customer's specific problem.

The system will prompt representatives while they are helping customers. For example, it can enable the FedEx worker to give instructions to the caller on how to measure the box he or she wants delivered, and even convert pounds to kilograms if necessary.

Senior staff at Federal Express are delighted with how EPSS has enabled the company to succeed in the rapidly changing global marketplace. According to senior management, EPSS has:

1. Helped improve the learner's job performance, not just her knowledge
2. Provided this help just in time, when and where the worker needs it
3. Furnished instant access to information, methods, tools, and decision aids
4. Used computer technology to leverage the expertise of a coach or mentor
5. Accelerated on-the-job training and retention of learning
6. Significantly reduced training time and cost
7. Increased flexibility with worker assignments
8. Enabled the organization to train difficult-to-reach workers
9. Decreased paper documentation, such as user manuals, evaluations, and tests
10. Increased employee self-sufficiency and empowerment

Technology-Based Learning — a Winner for Federal Express

Federal Express has invested large amounts of money in technology-based learning, but the company is quick to highlight the many benefits and even greater savings for the company. Internal studies at Federal Express have shown that its system for just-in-time training works. Instruction time on some modules has been reduced by 50% with no loss in retention or quality of training. Since the implementation of interactive video training, job knowledge test scores have increased an average of 20 points. Locations that have higher usage of interactive video training have higher job-knowledge test scores. When correlating test scores and performance evaluation ratings, Federal Express learned that, in general, the employees who have the highest scores on the test are, in fact, the company's better performers.

Federal Express firmly believes that its philosophy of "train to the job, perform to standards, and test for competency" provides customers with a value-added insurance program that translates into outstanding service and a competitive edge. A well-trained, knowledgeable, and empowered employee supports this philosophy and the company's goal of 100% customer satisfaction.

Ford Motor Company

Ford Motor Company is the world's largest producer of trucks and second largest producer of cars and trucks combined, selling over 7 million vehicles per year. Worldwide, Ford employs nearly 350,000 people and is the market leader in Britain, Spain, Australia, and Taiwan. Earnings exceeded $6 billion in 1997. Ford recognizes that creating customer value starts with delivering value, and high value starts with highly skilled workers and dealers.

Goal 2000 Demands New and Faster Learning

Recently Ford Motor Company established a challenging goal for itself: to become the world's leading auto maker by the year 2000. One of the ways in which Ford is seeking to accomplish this is to launch a record number of new or significantly improved vehicles. In the past two years, 18 new cars and trucks have been introduced. In fact, new models have been rolling off the assembly lines faster than Ford's training division could provide instruction on these models to service technicians, sales, marketing, and other dealer employees. A further complication in trying to keep its dealers trained and updated was the fact that over one third of all dealers were more than 100 miles away from one of Ford's 50 training centers, thus making it difficult, time-consuming, and costly for service technicians to be sent for training (Ruber).

"Automotive engineering has become so complex that it is impossible to get the job done with group-based classroom training anymore," says Ford's Larry Conley. He and other senior Ford managers acknowledge the reality that if Ford was to have its cars accepted by the general public, it needed a sales force that was well informed and technicians who were highly skilled and experienced in the new automotive technology. Dealerships could not afford to have poorly trained staff if they wanted to compete in the marketplace.

Faced with the paradoxical objective of reducing the cost of dealer training and communications while providing more learning within a shorter training cycle, Ford elected to implement its own learning information superhighway — the FORDSTAR Communications Network. FORDSTAR is a high-tech, innovative learning system that is the "highest-capacity, privately owned, satellite-based telecommunications network in the world."

Fordstar — Ford's Learning Technology Network

FORDSTAR was initially created over 10 years ago to improve internal communications with Ford employees. Today it broadcasts to over 250 regional sales offices and is believed to be the most comprehensive internal daily television news service in the world. FORDSTAR does much more; it has become Ford's learning network.

Ford has spent over $100 million to build this all-digital satellite communications network for training its dealers' employees throughout the U.S., Canada, and Mexico. While the idea of company-to-dealer television networks is not new to the automotive industry, no other company is equipped to send more than one television signal at a time; none uses high-speed compressed digital video technology; and none is equipped to use interactive voice and data as a basis for training and communication. FORDSTAR is capable of broadcasting up to eight video channels simultaneously on a single transponder (i.e., eight different training courses can be delivered at the same time) and has a transponder dedicated to data transmissions

Fordstar's Facilities and Resources

The classrooms are equipped with a television, satellite reception equipment, and interactive keypads used by the students to call the instructor and answer multiple choice, yes/no, and numeric questions (Figure 18.1). There are also student interaction computers that monitor the sites and capture the questions and quiz results from each student attending the broadcasts (Williams and Stahl). The Dearborn, Michigan, learning center provides eight channels of training-oriented programming. Dealerships downlink the telecourses and learning materials, permitting technicians, sales people, and other personnel to learn on-site.

The instructor's desk and broadcast facility puts the instructor in complete control of the various multimedia sources available. It includes:

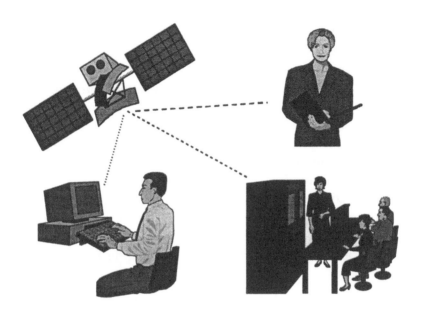

Figure 18.1   FORDSTAR's Learning Network

1. A standard computer for showing graphics and other computer-generated teaching aids
2. A laser disc player for showing videos and other graphics
3. An overhead camera for showing props and other objects
4. A switching console used to display various tools
5. A telestrator pen used to write on the screen
6. A camera that focuses solely on the instructor

On a monthly basis, Ford publishes a *Star Guide* for the dealerships, listing course descriptions and noting the intended audience. Attention is given to the different time zones, making it easy for dealers to decide which courses their employees should take and at what times. Course topics range from new model overviews and warranty and policy administration to anti-lock brakes and service department town hall meetings.

On-Site Coordination

An on-site facilitator serves several important roles in implementing the learning programs of FORDSTAR, including:

1. Acting as the contact person to whom the instructor can send the course materials
2. Receiving and distributing materials to each participant at his or her location
3. Administering and grading tests and reviewing results with students
4. Communicating with instructors about any issues or concerns; summarizing the test results. The summary sheets enable the instructor to tally the incorrect answers and review the most difficult problems on the air during the next broadcast — while the test is fresh in the students' memories.
5. Conducting orientation sessions to acquaint students with the technology and methods the participants and instructors will use to interact. A discussion of the technology and how it operates enables the participants to create a mental picture of how the learning will occur.

Interactivity — Key to Fordstar's Success

FORDSTAR utilizes an interactive distance-learning (IDL) system that features an automated, multifunctional control desk. To maximize interactivity, the control desk ("multimedia interactive platform" or MIP) is operated by an instructor, and the interactive response keypads allow continuous interaction between the instructor and learners. According to Conley, FORDSTAR's programs are designed to "call for some form of interactivity every 10 minutes."

This interactive capability is a key feature contributing to FORDSTAR's success. Here's how interactivity works:

1. Using response keypads, dealer employees log-in at the beginning of a class. This provides Ford and the instructors with an automated attendance record.
2. During the class, instructors encourage interactivity by issuing preformatted questions to the students from the host station.
3. The employees see the individual question on the TV monitor and the possible choices reflected on their keypad display.
4. When the instructor receives the results of the cumulative student population, he or she can broadcast these results back to the audience. Once the correct answer is highlighted, these dealers can gauge their progress in relation to others in classrooms throughout the network.

5. In addition to monitoring the progress of the entire class, instructors can also monitor each individual student's achievement. Each student's response to every question is logged on the instructor's computer, making the system an effective tool for courses that require certification or grading. The individual student performance records enable instructors to offer additional instruction to students who need help in specific areas, ensuring overall success.

Laura Thurman, consultant to FORDSTAR, notes how the keypads also "provide data feedback from the dealerships, creating a valuable audio link between the students and the instructor." If students have questions or comments, they can press the call button on their keypads to electronically raise their hand. The instructor in Dearborn recognizes callers by name and location within seconds and activates student microphones using the system's instructor touch screen. The ability to have two students on air at once facilitates interaction between the sites.

Improved Delivery and Performance at Less Cost

Conley reports that instructors are able to cover 20 to 30% more material during a telecourse, student grades are 20% better, and time away from dealership premises has been reduced by as much as 50%. Through FORDSTAR, the HRD staff is now able to reach its more than 45,000 technicians and 50,000 sales persons in a matter of days. The learning technology utilized has cut classroom training from an average of 31 days to just 9 days. Larger numbers of people can be trained simultaneously.

There is also a much higher quality of performance in the classroom because establishing a disciplined and consistent learning process is easier when information comes from a single course rather than from numerous courses throughout the country.

In addition, instructors tend to be more thoroughly organized when designing for distance learning because of the high expectations and high initial costs. It is wise to optimize the use of distance-learning technology because one is preparing for a large number of learners who will be participating at a single time. To gain a sense of the numbers — between January and October 1997, FORDSTAR delivered 4277 courses, totaling 11,968 broadcast hours. Attendance was over 300,000 logged-in students (Filipczak).

Another benefit is that since courses are being delivered by multiskilled instructional teams, the leader-experts in particular fields can be accessible to everyone in the satellite's footprint, while access to other instructors is available both during and after a course.

Students also tend to be more prepared and participative. Instead of experiencing several hours of course work and lectures at a single time, they can learn in a more paced and incremental manner over several sessions because they can learn gradually in the convenience of their offices rather than cramming it in during a week away from the dealership.

Electronic distribution reduces the cost of course materials. Utilization of satellite networks and CD-ROMs decreases per-student costs of faculty and facilities. Conley notes that an automotive electronics course for all of Ford's dealer technicians, which might cost $3.5 million to provide within classrooms, costs only $1 million to produce on CD-ROM.

Future Worldwide Expansion

Ford is now looking to expand into South America, Europe, and Australia. To spread distance learning to these sites more quickly, Ford plans to expand training via the network to internal employees at manufacturing and design plants. Learning through technology has indeed lifted learning, performance, and corporate success to world-class heights for Ford Motor Company.

Nortel

Nortel (Northern Telecom, Ltd.) is one of the world's most broadly diversified developers of telecommunications products and systems. A rapidly growing global telecommunications corporation, Nortel has over 63,000 employees in 200 locations around the world. Nortel's principal products are switching networks, enterprise networks, digital cellular systems, and high-capacity transmission equipment. It has quickly become a leader in developing products to help cable television operators and telephone companies deliver voice, data, and video services. Net income is over $15 billion a year, 10% of which is invested in research and development.

Nortel Learning Institute

To stay at the leading edge of technological advances, Nortel also spends an astonishing 10% of payroll costs in employee development activities.

State-of-the-art training facilities have been built worldwide, many of which work in partnership with local colleges and universities. Distance learning is becoming more and more central to Nortel's training efforts, much of it done through the Nortel Learning Institute.

Distance learning presents both challenges and opportunities for training of Nortel employees. The Nortel Learning Institute has been carefully analyzing how to utilize learning technology to efficiently and effectively deliver learning programs to employees located in every corner of the globe. When combined with other performance support tools, the Institute has determined that interactive distance learning (IDL) is the most productive path to pursue in building a global learning organization.

The infrastructure for training at the Nortel Learning Institute includes 1) the internal telecommunications network (both voice and data), and 2) the Nortel Vision Interactive Video Network. The Institute provides programs using CD-ROM, broadcast quality video, audioconferencing, and intranet applications. The three primary computer platforms used at Nortel are IBM PC or compatible, Macintosh personal computers, and UNIX-based workstations.

Audioconferencing with File Server/Intranet Access

Several of Nortel's training programs employ audio bridges to conduct the training session with participants located throughout the world. In addition to accessing the audio bridge, the participants are also expected to access files that exist on the file server or on the intranet. Nortel's experience is that training programs offered in this format are most successful when they last less than four hours because of learner and facilitator fatigue (Maxwell). The training programs are therefore scheduled over a period of several days or weeks. Like the Ford dealers mentioned earlier in this chapter, the participants appreciate that the training is delivered to their desks so they are not out for an extended period of time.

Because of the growing use and success of the intranet, Nortel has moved participant files to the intranet for programs that use on-line information during the training sessions. Files can be saved in formats that can then be read independent of the platform.

CD-ROM Multimedia

The Nortel New Employee Program is taken by all employees within six months of entering the company. Originally 4.5 days in length, it was redesigned to two days of classroom delivery plus a self-study CD-ROM for new employees to complete outside the classroom. Randy Maxwell, HRD Manager for the Nortel Learning Institute, states that the savings, using this format, have been substantial in terms of reduced costs for facilitators, for use of classrooms, and for employees' time. The Nortel Learning Institute has discovered, however, that the CD-ROM technology is less successful in Latin America and Asia, where learning cultures seek more social interaction. As a result, an alternative New Employee Program has been developed that includes more classroom time and a modification of the CD-ROM content for these cultures (see Chapter 17 for the impact of culture on the use of technology).

Interactive Video Network

The Nortel Vision Interactive Video Network can provide two-way video and audio at all the major North American sites, while one-way video, supporting return audio or key response systems for participant interaction is available for the other Nortel sites around the world.

Training sessions are generally limited to four hours or less to accommodate participants' attention spans, broadcast costs, and time zone differences. Nortel has discovered that the facilitator's role in video delivery is much different from traditional classroom delivery. Facilitators are coached in pronunciation, clarity of speech, and video IDL classroom participant management. There are also guidelines in terms of dress colors as well as graphic standards for presentation materials. Site facilitators are also used at the remote areas to increase participant interaction. In order to increase the interaction of participants with return audio, Nortel introduced competitive game activities among the sites. The game methodology has also allowed improved testing of the participants' level of understanding.

Web-Based Performance Support

Nortel has developed a set of performance competencies that have been moved from hard-copy handbook to the intranet in an interactive format. Connected with each competency is a list of suggested development activities, on-the-job applications, suggested readings, and recommended Nortel Learning Institute courses to support that competency.

The user performs the following steps in a typical session on the Nortel intranet:

1. Access handbook/tool through the Nortel Learning Institute web site
2. Enter anonymous user data
3. Complete self-assessment of the performance competencies critical to the job role (based on manager's feedback); learner identifies at one of the four operating levels of performance for each competency
4. Select development activities for application
5. Generate a personalized development plan
6. Save the file on hard drive
7. Integrate the development plan into career development activities

Participant and Facilitator Roles

Maxwell notes that the roles of the facilitator and participant change significantly when using distance-learning technologies, especially the balance needed between content presentation and participant interaction. To increase the interaction, Nortel uses extra facilitators and competitive type activities. Some of the most successful IDL broadcast projects have been developed with the sales and marketing teams because they often take on the role of entertainers. Since distance learners at Nortel are, for the most part, scientists and engineers who may resist IDL, Nortel uses technology that encourages both active participation and more self-directed learning.

Benefits of Course Delivered via Interactive Distance Learning

Staff at the Nortel Learning Institute are enthusiastic when describing the power and advantages of IDL. Maxwell cites the benefits Nortel gained as a result of a one-day Product Introduction course delivered via one-way satellite downlink:

1. *Speed of delivery* — The satellite course allowed 500 people to participate at one time. A classroom version would require 11 classroom days and would require approximately 30 calendar days to train the 500 participants. The inability and delay among Nortel workers to apply this product knowledge to the job would be much greater than the cost of technology.
2. *Visibility of change process* — The change is more visible when the initial change meeting involves 500+ people rather than being done with 15 to 20 people at a time.

3. *Opportunity for company-wide dialogue* — IDL provides opportunities for people to ask questions and hear from a wide variety of colleagues from different regional cultures and technical disciplines. The dimension of multiple perspectives enhances systems-wide thinking.

4. *Possibilities for large-scale organizational change* — Large numbers of people are allowed to interact with the ideas at the same time, reinforcing the viability of the change effort, and contributing to an accelerated implementation of the changes in systems and processes in the workplace. In addition, feedback and discussion from a large company-wide group provides for richer input into the design and delivery of the course, and results in excellent "double-loop" learning.

5. *Assessment* — Use of automated registration and keypads for feedback and evaluation allows for quicker, more comprehensive assessment of the effectiveness of the training.

Power and Impact of Learning Technology

Federal Express, Ford, and Nortel are continuing to expand their use of learning technologies to strengthen the power of their people and their organizations. Their leaders attribute much of their recent success to the quality and speed with which their workers have learned. Learning better and faster, however, is just one of the powers of technology. Being able to create and manage more and better knowledge is technology's other great contribution to success in the workplace. In Chapter 19, we will look at the organizations who are leaders in knowledge management technologies.

References

Filipczak, R. (December, 1997) Ford delivers on its satellite training — and plans more. *Technology for Learning*.

Maxwell, R. (1997) Distance learning and human development at Nortel: the shift from classrooms to communities of learning. Unpublished manuscript.

Ruber, P. (June, 1996) The grounds for training. *Beyond Computing*.

Thurman, L. (February, 1996) Ford takes the fast lane to dealer communication and training. *ED Journal*.

Williams, D. and Stahl, S. (November/December, 1996) Ford's lessons in distance learning. *Technical & Skills Training*.

Wilson, W. (June, 1994) Video training and testing supports customer service goals. *Personnel Journal*.

19 Best Practices in Knowledge Management Technologies

In Part III (Chapters 13 through 15), we explored the key technologies available for managing a company's intellectual assets, with particular emphasis on the use of the Web (Internet, intranet, LANs/WANs), Electronic Performance Support Systems, and Knowledge Engineering. More and more organizations are beginning to realize that the most important value of technology is its ability and power to acquire, store, analyze, and transfer corporate knowledge and that this corporate knowledge is crucial for providing customer satisfaction and corporate success. In this chapter, we will highlight four organizations that are at the forefront of using technology for managing knowledge — AMS, McKinsey & Company, National Semiconductor, and Price Waterhouse.

AMS

Headquartered in Fairfax, Virginia, AMS is an international business and information technology consulting firm which employs over 7000 employees worldwide. Founded in 1970, the company has shown profits for 27 consecutive years, with earnings of $1 billion in 1997. Priding itself as a leader in applying the power of information technology, AMS takes a holistic approach in linking people, processes, projects, and functions with technology to achieve "breakthrough performance" for each client.

AMS Center for Advanced Technologies (AMSCAT)

A major part of AMS's success is attributed to its Center for Advanced Technologies (AMSCAT), which includes a network of over 600 technologists who continuously create and share information and expertise. Let's look at how the Center assists in managing knowledge for AMS and its clients.

AMSCAT's mission is "putting innovative technologies to practical use in client environments." A number of collaborative research programs are ongoing throughout AMS to increase the breadth and depth of research across all industry practices, to research technologies of mutual interest, and broadly share that knowledge. Sharing knowledge is done through such means as publications, databases, and seminars such as "Technical Architects Meetings."

Over the past several years, AMSCAT has developed and shared a number of knowledge management technologies with its worldwide clients. Discussions of these follow.

Web Technologies

A key question faced by many of AMS' clients is: How to evolve an information systems infrastructure to leverage Internet technologies?" A few of the areas being explored and developed by AMS include:

1. The architectural complexity of connecting customers, customer service personnel, and legacy and external systems through the Web; this involves the study of the underlying technical issues of the Internet, intranets, protocols, and architectures
2. Security and technologies such as cryptographic techniques that will make electronic commerce possible over the Internet
3. Exploration of the strengths and weaknesses of Web development environments such as Java, JavaScript, plug-ins, CGI, and ActiveX.

AMSCAT recently applied this knowledge management technology in a British bank in which customers can apply for a mortgage over the Web and receive a response within minutes. This shows how a business can reach out to a large base of potential customers and forge close, service-oriented relationships with them.

The bank is also able to leverage the data in its back-end systems via Web-based applications. The bank now not only provides on-line decisioning services, but also tailors cross-selling strategies for related products to specific customers on the basis of profile and historical data from other bank activities.

Data Mining

Over the past 30 years, organizations have become skilled at capturing and storing large amounts of operational data. Unfortunately, we have not seen corresponding advances in techniques to analyze this data. Until recently, manual analysis with report and query tools was the norm, but this approach fails as the volume and dimensionality of the data increase. New approaches and tools are needed to analyze very large databases and "mine" their contents. Data mining enables organizations to find meaning in often overwhelming and confusing data. By discovering new patterns or fitting models to the data, professionals can extract information to develop strategies and answer complex business questions.

AMS has taken a leadership role in the field of data mining which involves 1) navigating data, 2) discovering patterns in the data and creating new strategies, 3) using underlying statistical and quantitative methods of visualization, 4) employing platforms to support the tools, 5) identifying the techniques to prepare the data, and 6) finding ways to quantify the results.

Visualization tools used by AMS include AVS/Express, SGI MineSet, and Visible Decisions Discovery. Integration of data mining products is undertaken by tools such as DataMind and IBM's Intelligent Miner, and research technologies are utilized such as neural networks, decision trees, clustering, association, and rule induction.

AMS and their banking client in the U.K. used data mining techniques to analyze more than 267,000 records of 132,000 customers to find correlations among customer behavior, demographics, and profitability. To look at these large volumes of data across 152 business dimensions, AMS used sophisticated statistical and visualization techniques. These included scatter plot matrices and parallel coordinate plots in conjunction with brushing, cutting, and high-dimension rotation.

Applying these techniques provided AMS and the banking analysts with new perspectives on their data. They rotated and twisted a multidimensional image to search for relationships among groups of data to discover new patterns or trends. In this case, AMS was able to find correlations between channel usage (e.g., checks, branch visits, ATM use), age, and occupation to help the bank quickly zero in on the most profitable segments of bank customers.

Team Learning Technologies

In today's global workplace, enterprises increasingly rely on project teams whose members are dispersed around the globe. Technology, especially communications networks, extends the enterprise beyond its traditional walls and enables these organizations to share information, resources and knowledge. Working (and learning) together requires the effective use of collaborative work technologies. AMS has employed intranet groupware tools such as Netscape Collabra and Lotus Domino, desktop conferencing products such as White Pine's CU-SeeMe and Intel's ProShare.

Electronic Commerce

From sources such as electronic mail and on-line catalogues, enterprises seek to leverage the opportunities of electronic commerce. Businesses that provide multiple channels are likely to be more successful because they offer customers the following:

1. Choice — ways to interact with the organization
2. Convenience — the most appropriate channel at different times and places
3. Customization — tailor goods and services to individuals through technologies such as interactive voice response (IVR) menus that put customer's usual choice first, or Web pages that adapt to the customer's browser and line speed

Bridging separate systems via transactional processing middleware ("software that enables distributed application components to interoperate across networks despite differences in communications protocols, hardware platforms, operating systems, databases, and other application services. AMS uses 1) Web/Java connectivity tools to relational and object databases, 2) Web/Java gateways to middleware (like MQSeries and Tuxedo) and back-end systems, 3) Java-to-Java-communications using remote method invocation and object streaming, 4) publish-and-subscribe middleware interfacing with Java, object and Web technologies, and 5) the integration of Web and Object Request Brokers (e.g., VisiBroker) to determine which products best increase the efficiency of data transfer, standardize processing requests, and integrate existing agency systems.

Database Management

Like other global companies competing in the global marketplace, AMS needs to store and manage large amounts of data. AMS has examined the impact of advances in data-based technologies on both on-line transaction processing and systems and data warehouses. In 1996 AMS began building a large data warehouse in conjunction with its Federal Financial System. The initial implementation is a 20 GB Oracle database on their Pyramid Platform.

Integration of Work and Learning

To remain competitive, AMS recognizes that it must provide the framework and means that allows staff to stay ahead of rapid change, while "continuously improving its ability to create value and achieve results." To accomplish this, a new approach to accelerate individual, team, and organizational learning and development called "Integration of Work and Learning" has been initiated. According to Lew Parks, VP of Human Resources, the primary context of learning in this approach is "the assignment and the way in which it maps to learning." This approach aligns all of AMS learning and development resources to workplace activities and brings resources to "bear as close to the time of need as possible."

An example of this focus is Competency Based Career Development, which links employees to the assignment-as-learning experience and to the resources that support their roles. This approach enables workers to better understand what is expected of them as learners, assess where their learning gaps exist in order to fulfill their role, and access the resources they need on a timely basis.

When they have completed their assignments, AMS associates can evaluate 1) what they have learned, 2) how it has contributed to their growth and development, and 3) how they will share their knowledge with others. They are able to manage the next step in the assignment process because they are "aware of the competencies needed to achieve results in subsequent assignments and the learning resources required to get there" (Parks).

AMS believes that its new educational approach achieves results for several reasons:

1. It employs problem-based learning through using actual work experiences.
2. It uses work activities and tools on-the-job and in the classroom.

3. It delivers education just in time, just enough, and at the point of need.
4. It supports individual and career development needs.
5. It shortens the learning cycle and improves learning retention.

McKinsey & Company

McKinsey & Company is an international management consulting firm that advises the top management of companies around the world on issues of strategy, organization development, and operations. Tom Peters, Bob Weatherman, and Kenichi Ohmae are just a few of McKinsey's long list of past and present associates.

McKinsey is making extensive use of technology to manage its corporate knowledge. They have developed 31 practice information centers — 18 centers of competence for functional specialties like marketing and organizational performance, and 13 for industries like banking, insurance, energy, and electronics. At McKinsey, there is an emphasis on the systematic development of consultant skills, so that the firm will have as much "internal knowledge on demand" as possible. The organization has created the ability to tap systematically into what people have learned.

Customer Service via Knowledge Management

The storing and sharing of knowledge within and outside McKinsey has greatly improved the quality of customer service. In the past, McKinsey associates had to rely on their personal network of contacts when meeting client requests for information and/or help which fell outside their own individual realm of expertise. The sources for new information were limited by the size of a consultant's personal network.

By electronically linking every associate together with LANs and WANs, and at the same time developing knowledge databases containing previous consulting experiences, industry information, and expert contacts, the number of sources for new information available to any single McKinsey associate has been greatly increased. Additionally, discussion databases organized around specific topics has allowed ongoing knowledge sharing between geographically distant employees of the firms. Anyone at anytime in any location could monitor ongoing discussions, take away new insights, or join in and add new knowledge throughout the computer-assisted learning system.

Knowledge Management Strategies

All too often, valuable lessons, whether from successes or heart-wrenching failures, never leave the minds of the individual or group involved. Learning organizations know how to capture these lessons through a variety of positive, incentive-based methods. To optimize the capturing of McKinsey knowledge, workers are rewarded for putting their learning into databases in the practice information centers. Consider some of the knowledge management strategies used by McKinsey:

1. A Director of Knowledge Management has been appointed to coordinate company efforts in creating and collecting knowledge. (Philip Morris calls the position "Knowledge Champion;" Dow Chemical has a "Director of Intellectual Asset Management.")
2. Knowledge transfer is seen as a professional responsibility and part of everyone's job.
3. Knowledge development is included in the personnel evaluation process.
4. An employee does not get a billing code (and therefore reimbursement) until he or she has prepared a two-page summary of how and what he or she has learned from the project.
5. Every three months, each project manager receives a printout of what he or she has put into the company's Practice Information System.
6. An on-line information system called the Practice Development Network (PDNet) is updated weekly and now has over 6000 documents. Documentation also includes the Knowledge Resource Directory (McKinsey's Yellow Pages) that provides a guide to "who knows what."
7. For any of McKinsey's 31 Practice Areas, one will be able to find the list of members, experts, and core documents.
8. A McKinsey Center *Bulletin* appears two or three times per week for each of the Practice areas, featuring new ideas and information that a particular Practice area wants to "parade" in front of all the company's staff (Peters).

Based on their experience in knowledge management, McKinsey offers the following advice:

1. Knowledge-based strategies must begin with strategy, not knowledge or technology.
2. Strategies need to be linked to organizational performance and success.

3. Executing a knowledge-based strategy is not about managing knowledge, but nurturing people with knowledge.
4. Organizations leverage knowledge through networks of people who collaborate, not through networks of technologies that interconnect.

National Semiconductor

National Semiconductor prides itself as a company that creates technologies for "moving and shaping information" by manufacturing products that connect people to electronics and electronic networks. Market segments include business communication, personal computing, automotive, and consumer audio, and products such as personal computers, cordless telephones, computer security, auto instrumentation, multimedia centers, and microwave ovens. Customers include many of the global 1000 — AT&T, Siemens, Intel, Apple, IBM, Boeing, Sony, Toshiba, and Ford, among others.

A pioneer in the semiconductor industry, National Semiconductor was established in 1959. Since that time, the company has been at the vanguard of revolutionary electronics technologies, and is today an acknowledged leader in the design and manufacture of the products that provide access to the information highway. Corporate sales in 1997 topped $2.5 billion. Over 12,000 employees work at manufacturing sites in Asia, Europe, and North America.

Leadership in Integrating Analog and Digital Technologies

National's mission is to deliver to its customers complete single-chip systems that are integrated, tested, and shipped ready-to-use. To achieve this, the company has committed itself to three strategic imperatives:

- State of the art process technology
- World-class manufacturing
- Six-month time-to-market methodology

Obviously, managing knowledge is critical for National Semiconductor. Let's examine how the company does it.

Knowledge Storage

Like many organizations, National was overwhelmed with information, retaining data that was not needed and losing valuable data that should have been stored. Employees were uncertain as to what to remember and what to leave behind. In order to cure its "corporate amnesia," the company began developing a corporate database and storage system. Here's how the knowledge management system was described in a recent issue of the company newsletter, the *InterNational News*:

> Imagine a city block filled with many different libraries. Stacked on those miles of shelves are millions of books and articles, tons of art and graphics, even videotape, film, and audio recordings. Now imagine that you want to seek information, but you have only a single library card and can enter just one of those libraries. Worse yet, you have heard that other, better libraries exist in other cities around the world, but you have no idea where they are or what's in them. Frustrating, isn't it? That was the knowledge storage and access situation at National until recently. Now mountains of information stored in different computer systems can be made available to staff throughout the company via a new project called "Knowledge at the Desktop."

Employees at National now have access to all kinds of information, such as business plans, materials data, customer support, field sales, public relations, and human resources data. They also have access information from media sources, Sematech, news wires, patent and research records, and bibliographies. Not only text, but graphics, sound, and even video will be available. Fulcrum, a sophisticated new computer program, enables employees access information on any National system, from whatever type of terminal or workstation is at one's desk. A special corporate action team has successfully worked to "create an easy method for people to find their way through the information jungle," according to Mary Holland, Manager of the National Technical Library.

National Semiconductor cites three key benefits from its knowledge-sharing technology:

1. Increased participation from every part of the organization
2. Automatic knowledge storage and retrieval as well as corporate memory
3. Wider understanding and shared interpretations of knowledge bases (Miles)

Sharing Technology

National Semiconductor holds annual in-house International Technology and Innovation Conferences for the purpose of encouraging their technologists to create and share core technologies to "further National's *Aspiration 2000* technology roadmap." Papers were recently presented in the following areas:

Circuit design and simulation
Manufacturing
Device technology and architecture
Process technology
Packaging
Test systems and methodologies
System and software development

Video Compression Technology for Learning

National Semiconductor has invested intensive creativity, time, and research in the development of video technology. Why is video compression so important to National? Because the interactive video capabilities of the new products such as document-sharing in real-time would not be possible without video compression — the available bandwidth of telephones is too narrow to send real-time motion video, and video data take up too much storage space. Video compression is a critical innovation for video playback, video conferencing, and video phones.

With this technology, National is able to develop a document or share information with co-workers halfway around the world in real-time. The data is entered on a PC and appears instantly on their screens. The team members receive feedback immediately. Together they can finalize specifications for a new run of chips or create a proposal for a new customer more quickly and conveniently than was ever thought possible.

Price Waterhouse

Price Waterhouse, a global consulting firm headquartered in New York, estimates that up to 99% of its revenue is generated from knowledge-based professional services and knowledge-based products. Its 56,000 employees located in over 400 offices worldwide spend up to 80 hours a year "creating and sharing knowledge" (APQC).

Why is knowledge management so critical to Price Waterhouse? Among the driving forces are:

1. The geographic dispersion and need to serve global clients requires a timely coordination and sharing of information between the different subsidiaries.
2. Information about tax policies, legislation, banking, financial requirement, etc., changes so rapidly that it needs to be captured and disseminated quickly.
3. Employee turnover of 15 to 25% per year necessitates that systems capture this knowledge or it will be lost.
4. It is important to discover and constantly share best practices throughout the firm.
5. Employees must know about former projects and projects conducted in other parts of the clients' organization.
6. The tendency to have "islands of knowledge activity" creates inefficiencies and duplication.

Knowledge Management Resources

To meet the growing demands for acquiring, analyzing, and storing knowledge, Price Waterhouse has created a variety of Knowledge Management resources:

- *KnowledgeView* — PW's best practices program (discussed below)
- *Change Integration* — A business line for change management and re-engineering (which represents 40% of PW's consulting business)
- *Information Technology* — External knowledge support for PW clients
- *Advanced Systems Engineering Centers* — Includes six software centers in the U.S. and two in Europe
- *Information Specialists*
- *Centers of Excellence*
- *Industry-related knowledge* (e.g., petroleum) and *process-related knowledge* (e.g., supply chain management)

KnowledgeView

KnowledgeView, PW's proprietary best practices repository of information, has been gathered from more than 2200 companies worldwide and contains more than 4500 entries, with references to more than 350 internal and external benchmarking studies. The goal of KnowledgeView is to "support the firm's core competency of being business advisors: including the accumulation, analysis, creation, and dissemination of value-added information and knowledge that PW professionals can use to improve business performance of clients, and ultimately increase the value of PW's services" (APQC).

KnowledgeView is Lotus-Notes-based rather than CD-ROM-based, thereby allowing daily information updates as well as the capacity to access and share knowledge "instantaneously on a worldwide basis." The databases in KnowledgeView incorporate the following information:

1. Best practices as identified in PW and non-PW programs
2. Benchmarking studies from internal and external sources
3. Expert opinions synthesized by industry or process subject matter experts
4. Abstracts of books and articles on business improvement
5. PW staff biographies with resumes database according to country, industry, skill, language, etc.
6. Views and forecasts of PW's own experts on important topics

KnowledgeView is classified according to industry, process, enabler, topic and measurement so that PW consultants can target and find the knowledge for which they are searching. An important feature of KnowledgeView is the format used for containing the information. For example, in the "best practices" database, the format is established to answer the following questions:

- What caused the change?
- What old process needed improvement?
- What is the new process?
- What is the new performance and how is it measured?
- What were the lessons learned?
- What are the future directions?

KnowledgeView is maintained and updated at four Knowledge Centers located in Dallas, London, Sydney, and Sao Paulo. Through these regional centers, local staff have more immediate access points as well as help in stimulating knowledge sharing.

In 1996 Price Waterhouse received the "Best Practices Award" from *PC Week* magazine for its KnowledgeView technology.

Successes and Learning from Knowledge Management at Price Waterhouse

Price Waterhouse has learned and gained tremendously as a result of its various knowledge management efforts. PW cites the following as key lessons learned:

1. It is important for all PW people to fully understand and value knowledge sharing and knowledge creation.
2. One should add knowledge sharing to every situation.
3. PW should tie knowledge management directly to strategy for every program.
4. Knowledge management must be continually "fueled and regenerated."
5. It is important to standardize knowledge management architecture and hardware.
6. PW must systematize the knowledge management infrastructure.

Based on its innovative management of knowledge in five different areas — strategy, approaches and processes, culture, technology, and measurement — Price Waterhouse was recently selected as an Emerging Best Practices Company in knowledge management by the International Benchmarking Clearinghouse (IBC).

Technology — the Essential Tool for Managing Intellectual Assets

AMS, McKinsey & Company, National Semiconductor, and Price Waterhouse are just a few among the emerging group of companies that are incorporating knowledge-management technologies into the lifeline of their strategies and operations. Many other firms (e.g., Chevron, Texas Instruments, Ernst & Young, Canadian Bank of Imperial Commerce, Arthur Andersen, Skandia, Texas Instruments, Dow Chemical, Monsanto) have vigorously invested in, developed, and trained employees in knowledge management technology and have developed extensive knowledge management systems (Allee). They have recognized the importance and truth of what Walter Wriston, former CEO of Citibank, recently stated: "The organization that can harness its knowledge will blow the competition away!"

References

Allee, V. (November, 1997) 12 principles of knowledge management. *Training and Development.*

American Productivity and Quality Center (1998). Company Case Studies: Price Waterhouse. Unpublished.

Marquardt, M. (1996) *Building the Learning Organization.* New York: McGraw Hill.

Miles, R. (1997) *Corporate Comeback.* San Francisco: Jossey-Bass.

Parks, Lew. (1997) A new approach to education accelerates learning and development at AMS. *Making Connections.*

Peters, T. (1992) *Liberation Management.* New York: Alfred A. Knopf.

20 The Future of Technology in the Workplace

Thus far in this book we have described the operations and impact of technology in the workplace, ranging from traditional applications such as television or computer networks to emerging developments such as EPSS and data mining. We have noted how technology changes the nature of individual and organizational learning and documented these changes with numerous examples and case studies.

In this final chapter, we will move from present realities and possibilities to speculate on how technology will affect the future workplace. Of course, predicting the future, particularly in these days of turbocharged change, is risky. These scenarios, however, depict exciting new ways in which technology could be used in the future for learning and managing knowledge in the workplace. The people and organizations mentioned in these scenarios are fictitious, but they are reasonable extrapolations of what is taking place today on the leading edge of technology use around the world. How accurately do they portray the future? We'll have to wait a decade or two to find out!

Key Largo Learning Associates — a Virtual Teaching Organization

Historically, institutions of specialized and higher learning have had a major physical presence: buildings, classrooms, libraries, etc. In the case of colleges and universities, this means a campus with many facilities, including gyms, cafeterias, dormitories, parking lots, and so on. In terms of evaluating the quality of an institution, these physical facilities are often important factors that have to be taken into account, even though they typically contribute little to the academic or research outcomes of the institution. The quality of an educational institution has often been synonymous with resources; so the more facilities, the more likely the educational institution was seen to be better equipped for academic or research activities. But in the information age, this assumption is not really valid any more.

Key Largo Learning Associates (KLLA) is located in Key Largo, Florida, because that's where the six principals of the organization like to get together in the winter for fishing expeditions. Indeed, there is a small office there with three support staff who handle all administrative matters. But most of the time, the six "professors" are in other parts of the world consulting, teaching, or attending conferences. They are all experts in genetics and bioengineering who taught at traditional universities for many years but decided they could accomplish more with greater flexibility by forming their own "virtual college" in the form of a small partnership, similar to a medical or law practice. The partnership (which is non-profit) is accredited by the regional accrediting agency and in most respects behaves like a department in a small college — without any physical facilities. They offer about a dozen different courses in genetics and bioengineering to students from around the world, with all interaction being done electronically. Some students are enrolled in degree programs at other universities, but many are individuals who work for companies or are conducting research in government labs. Students can take individual courses or complete six courses for a certificate.

All of the facilities required for academic and research functions are available on-line, including journals, books, and conference proceedings. So there is no need for a physical library. Similarly, the various computing resources needed for research and lab sessions are available on machines (including a number of supercomputers) around the country, which KLLA pays to use. Much of the interaction with students, staff, and outside experts is done via desktop video, so there is plenty of "face-to-face" contact, even though nobody ever needs to travel anywhere. In addition, many of the teaching materials and actual course activities are openly available on the network, so the quality of the academic offerings is easy to scrutinize and evaluate. Indeed, for every student formally enrolled in KLLA courses, there are dozens of "observers" who benefit from informal participation.

KLLA achieves its success (and there are always waiting lists for enrollment) through the quality of instruction and the expertise of the partners in the subject domain. All KLLA resources are devoted to functions essential to learning, teaching, or research. This raises the question: In the 21st century, will traditional colleges and universities become dinosaurs?

California Custom Car Manufacturing, Inc. — Consumer CAD/CAM

The automobile is a defining element of the 20th century, yet since their appearance at the beginning of the century, there has been one basic model for their manufacture: the assembly line and mass production. While a customer has a great deal of choice in terms of accessory options, the car itself is predefined. But with today's technology, it doesn't have to be this way.

California Custom Car Manufacturing (CCCM) Inc., located in Los Angeles, builds cars to individual tastes at a price equivalent to conventional mass-produced vehicles. The technology that makes this possible is Computer Aided Design/Computer Aided Manufacturing (CAD/CAM). This is intelligent CAD/CAM, capable of determining how different automobile components can (or cannot) fit together.

Customers sign on to the CCCM network and are guided through the car design process by an expert system that helps them decide what components (engines, body styles, seats, doors, instruments, etc.) fit their needs/wants. This is a "what-if" type of system that allows people to experiment with different ideas and "undo" or save a choice as they go along. The estimated final price of the current design is shown so the customers can assess the cost implications of their choices. The expert system also warns them about operating or maintenance considerations of the components they select. As soon as the car is "drivable," customers can take it on a virtual reality test drive to see how they like what they have designed so far.

Once customers are happy with their designs and decide to go ahead and purchase the car, they make electronic payment arrangements. When these are satisfactory, the order is submitted and manufacture of the car begins. The CAD program generates a complete parts list, which is then transmitted to the various parts suppliers who fill the order and have the parts shipped the next day. Once the parts arrive at the CCCM factory, they are assembled in a staging area for each vehicle (each part is bar-coded with the order number) and a small team of smart robots begins assembly according to the CAD specifications. Many QC checks are made electronically as the assembly progresses, and the completed vehicle is tested in a road simulator which checks all operational aspects. The final, approved car is then given a final inspection by a human inspector (the first and only involvement of an actual person in the entire process) and delivered directly to the customer's home via truck. CCCM promises delivery within seven days under normal circumstances. CCCM also completely guarantees all aspects of the car for 10 years — and has a higher reliability and customer satisfaction rating than traditional car companies. Also, each car is completely unique. Needless to say, CCCM is a very successful business.

CCCM demonstrates the incredible power of CAD/CAM coupled with intelligent (expert) systems. This raises the question: How viable will conventional mass production manufacturing enterprises be in the 21st century?

Ecofinance Associates — a Global Financial Services Partnership

One industry sector that has adopted technology very extensively is the financial services domain, including banking, insurance, stock brokerage, and credit card operations. Indeed, no aspect of financial services as we know them would be possible without sophisticated computer networks. Anyone who has ever used an ATM or credit card in a foreign city and gotten an instant verification, can't help being amazed at how efficiently the global financial system works.

EcoFinance Associates is a small group of 22 financial experts who live in different countries and provide a global service network for clients who wish to invest funds in ecologically worthwhile ventures. In addition to their regional knowledge, each partner has certain specializations and specific financial expertise. When a client (or prospective client) makes an inquiry about a possible investment, he or she is linked electronically to the partner who best matches that interest (a message is sent out to everyone via an e-mail list). A weekly staff meeting is conducted asynchronously over the course of a day using a conference system in which all partners summarize their current projects and share concerns. Small teams may work together on a particular portfolio, sometimes interacting via real-time videoconferences (desktop)…although differences in time zone limit the feasibility of this kind of communication.

What is especially interesting about the partners is the diversity of their life-styles and work habits. Some only work a couple of days per week or a few hours per day. Others work many hours over the course of a week or two on a specific portfolio and then take a week or two off. Some people devote their "free" time to recreational pursuits, their family, or other professional activities (e.g., writing, consulting, teaching). The only requirement is that everyone read their e-mail and conference postings regularly and participate in the weekly staff meeting. Since these are asynchronous events, the individuals can fulfill these minimal duties whenever they wish and, using laptop/hand-held computers, from where ever they desire. One partner works from his yacht equipped with a satellite transponder!

Management of the organization is very lean; it consists of just two people: the president and an administrative assistant. EcoFinance has a virtual headquarters maintained via e-mail, the web, and worldwide 800 phone/fax numbers. It has its own network server to support on-line operations (which is maintained and supported by a vendor). All administrative tasks (e.g., billing, accounting, payroll, etc.) are handled automatically by HRIS software, most of which is linked to transactions performed by the partners. The partners are responsible for taking care of their own hardware, software, and network connections, although support is available from their vendor if needed, and there is a lot of sharing of information across the partners.

The remarkable thing about EcoFinance is that it is extremely successful and has a very satisfied clientele. Because it can pool the expertise of some very knowledgeable individuals from around the world, the organization presents a great deal of depth to clients. Furthermore, the partners are able to establish a close working relationship through technology (e.g., e-mail, fax, phone) with their clients and be very responsive to their needs. Finally, the organization is able to accommodate the work styles of its partners and, hence, keep them motivated. Since they all share in the profits, they have a strong desire to help each other and work together.

United Clothing Suppliers — Weaving Things Together Electronically

Most large manufacturing/retail operations involve global operations in terms of where components originate, are assembled, and eventually sold. Nowhere is this more true than the clothing industry where the materials, design, manufacture, and selling are likely to transpire in very different parts of the world. However, in the 20th and previous centuries, this involved the physical transfer of goods from one place to another; but in the 21st century, it is mostly information about clothing that is transferred.

United Clothing Suppliers is an international conglomerate of 10 large apparel companies on different continents. While each company conducts its own retailing operations, they all participate in joint manufacturing and buying efforts. Indeed, one of the most profitable ways to make money in the fashion industry to take a style or trend that is successful in one region of the world and introduce it somewhere else. United is very good at doing this because it can quickly move a product line to market from one country to another through electronic resources. This speed makes the company very competitive.

Detailed product information, including patterns, specifications, photos and marketing footage (even fashion shows) are kept in on-line databases that can be accessed by any company in the conglomerate. The database also includes complete manufacturing records. Furthermore, detailed marketing data on the demographics of customers, including relevant advertising materials, are also available, so product planners in each company can determine how much it might cost them to make certain product lines and who they might sell to in their countries. They can also contact suppliers and ask them for manufacturing quotes by supplying them with the specifications available in the database.

While the database is very important, it is only one component of their electronic resources. Indeed, on-line interaction among product planners is just as vital. Every quarter, buyers and marketing staff from each of the companies participate in a virtual "show" in which current and new product lines are presented by designers around the world. These presentations are multimedia in nature and include photos, video clips, animations, and music. Buyers can discuss purchases via asynchronous or real-time videoconferencing, and there are many specialized conferences focused on particular market segments or clothing categories. During these quarterly shows, each company makes its buying decisions and lines up its suppliers and manufacturing facilities.

This mix of information databases and electronic conferencing, along with expert systems and EPSS, is typical of the knowledge management info-structures that will be employed by organizations in the 21st century to make them competitive and successful.

The Newtown Building Department — Computerized Government

While almost all government agencies are computerized, in many cases the automation of their activities seems less productive than older manual means of operation. It doesn't have to be that way, as the Newtown building department demonstrates in an aspect of local government that is notoriously slow, frustrating, and susceptible to being compromised.

All building permit applications are submitted electronically, including plans/drawings that must be in one of the dozen software formats that are accepted. Once the application is received, the applicant is automatically sent a confirmation, and the plans are analyzed by an expert system that checks them for compliance with all relevant codes and regulations for the location and nature of the structure. If any violations/problems are identified, a message is immediately sent back to the applicant, indicating the deviation and recommendations for addressing the problem. If applicants want to request an exception or argue their case, they can do so by posting messages to a conferencing system where they will get immediate response from building department employees; discussion can continue until the issue is resolved or the applicant requests a face-to-face hearing via desktop video. In the case of a large-scale development (e.g., a shopping center, office building, hospital, etc.), there would be many additional documents and meetings needed (such as environmental impact or traffic studies), and the application process triggers the scheduling of additional on-line discussions or meetings.

Once an application has been approved, a permit is generated along with an inspection schedule and transmitted to the applicant. As each inspection deadline occurs (or the applicant requests an earlier inspection), a confirmation of inspection is sent and an inspector is dispatched for the appropriate time and place. Once on-site, the inspector places various kinds of sensors on critical elements of the building (i.e., beams, walls, pipes, circuits, etc.) and takes readings using a hand-held computer. These are automatically transmitted back to the computer at the building department which checks them against the plans to ensure compliance. If they are acceptable, the approval is transmitted back to the inspector on-site who lets the builder know everything is OK. On the other hand, if there are problems, an explanation of these is transmitted, and the inspector provides them to the builder, withholding approval until the problems are rectified. In the case of a large-scale commercial development, the inspection information can be transmitted to the builder/contractor offices, where changes to the plans can be made on-line and then relayed to the building site.

While such an automated system is not without its headaches (think of the building bottlenecks when the system goes down), it could speed up the building permit and inspection process enormously, produce better quality buildings, and eliminate some of the oversights associated with a traditional building department. This raises the question: In the 21st century, can government agencies and their staffs function properly without learning how to use technology?

Greenville Grocery Mart — an On-Line Retail Environment

While many functions can be automated, eating is one human activity that cannot. On the other hand, shopping for food can be automated. Of course, many people find the process of visiting a grocery store enjoyable. But for some (the handicapped, seniors, people with young children in tow, etc.), it is not always a pleasant experience. In the case of commercial institutions, restaurants and cafeterias often require specialized foods or bulk orders that cannot be filled by ordinary grocery stores. Technology can allow small local retailers to offer the selection and service of a large nationwide chain, even though they have a fraction of the resources.

The Greenville Grocery Mart (GGM) addresses these concerns by providing an on-line grocery shopping environment (but no actual store people can visit). Customers sign on to the GGM system and create their order by selecting items using various directories, by traveling around a virtual store and viewing items on shelves (click and point shopping), or browsing through simulated meals/menus and having the item list generated for them by the system. The shopping/ordering interface can be configured by the customer to be very simple and fast — or quite intricate and involved. All previous orders are kept in the customers' files, so they simply repeat or modify a past order. For each item in their "store," full details about the ingredients, nutritional value, packaging, manufacturer, and recommended uses (including recipes) are provided.

Once an order is completed and electronic payment is authorized, it is transmitted to a large refrigerated warehouse where people (they do a better job in this task than robots and provide local employment) drive around in electric carts collecting all the items and delivering them to a packaging area. Each order is packaged in a delivery container and shipped by truck to the customer. At the customer's location, it is delivered as requested (e.g., the kitchen table, restaurant loading dock). Once the customer has accepted the delivery, the payment transaction is processed (less the price of any items that are not acceptable). If customers order items that are not in stock at the warehouse, they are notified of this when they are placing the order and if they still want the item, an order is transmitted to the vendor for next-day delivery to the GGM warehouse.

This is just the basic functionality of the on-line shopping environment. The system also includes discussion areas where customers can interact via real-time video with other shoppers or in asynchronous conferences devoted to every imaginable food-related topic. There are frequent cooking demonstrations in which vendors showcase their products and, while on-line customers can't taste the results, those who have smell-equipped systems can check out the aromas. There are also regular appearances by well-known chefs who talk about their cooking (and plug their latest cookbooks — available on-line, of course). These latter activities are sponsored by product vendors and typically done via real-time video so people can ask questions directly.

While on-line grocery shopping may not be for everybody, it meets the needs of many customers. In fact, it provides selection and offerings not possible in the traditional corner grocery store. This raises the question: In the 21st century, what difference does the size of the business make if small companies can offer products and services just as well as large ones?

The Neighborhood Clinic — Telemedicine and On-Line Health Care

Medicine and health care have always been dependent on technology — beginning with simple devices like the thermometer and stethoscope. Today, medical diagnostic equipment (e.g., CAT-scan, MNR, endoscope) is becoming increasingly sophisticated — but nothing like what the future holds! However, at present, only the larger clinics and hospitals can afford to have the best and most advanced equipment. This is likely to change.

Ironically, one of the major breakthroughs in future medical care is pretty mundane — the patient record card. This is a "smart card" resembling a credit card, that contains the individual's complete medical records and will likely be issued by health-care providers or insurers. When a person goes to a clinic, he or she passes the card through a device and all medical information is downloaded and immediately available to any member of the health-care team on their office, hand-held, or body display unit. As test results, diagnoses, or treatments are rendered, they are put into the record, which is updated to the card when the person checks out of the facility. Furthermore, individuals can view their records on their own computers anytime they wish.

To ensure the integrity of the data, there are various levels of encoding and security which only allow properly authorized individuals to enter information.

Most diagnostic work is done by expert systems. When test results and observations made by health-care providers are fed into these expert systems, they analyze the patient record along with trying to match symptom patterns against large clinical databases. They recommend treatments or further tests to the medical team. These expert systems are commercial products sold by health-care software companies that provide extensive evaluation reports supporting the accuracy and reliability of their systems (similar to the way drugs are marketed). Simpler, less expensive versions of the expert systems are available in the consumer marketplace for people to use at home as personal health-care tools.

Other important components of advanced medical systems are small, portable electronic sensor kits which can automatically measure vital signs (e.g., temperature, pulse, heart rate, etc.) as well as do more sophisticated things like ultrasound, MNR, and X-ray scans. The data collected by these kits are fed directly into the computer and expert diagnostic systems. This allows very rapid assessments in a health-care clinic as well as home or field settings.

Telemedicine allows an interesting development — the return of house calls. But these are house calls done by desktop video. The health-care provider links up with the patient at home or the office via desktop video and conducts an examination. If tests are needed, providers can ask the patient to go to the nearest neighborhood clinic where sensor kits are available and have the data collected for their perusal (or more likely, to be fed into a particular expert system). In the case of a chronically ill patient, the appropriate sensor kit might be provided to them at home, so the health-care provider can run tests any time by asking the patient to position the sensor in certain body locations. It would also be easy for a provider to bring a specialist or other person in to consult by asking him or her to link into the videoconference from a computer.

The most important characteristic about all these technology developments is that they will be sufficiently inexpensive that neighborhood clinics can provide high-quality care, which traditionally would only be found at the largest, best equipped medical centers. Indeed, individuals will be able to personally take advantage of many of these technologies on their own home computer systems. So the question for the 21st century is, how will technology and patient knowledge change the field of medicine?

Acme Printing Systems — Electronic Troubleshooting

The emergence of the information age has resulted in dramatic changes in the skills needed to operate and maintain all types of equipment and machinery. Equipment that was electromechanical in the 1950s became electronic and digital by the end of the century. Furthermore, most new equipment is designed to work in a network environment, i.e., to interact with other systems. This includes vehicles, office machines, appliances, weapons systems, and so on. So the kind of learning and ongoing training needed for those who work with equipment (a large segment of the work force) has altered significantly between the 20th and 21st centuries.

Chris has worked as a repair technician for Acme Printing Systems for about 10 years. In that time, Chris has seen the kinds of equipment sold and installed by systems change considerably, along with the nature of the maintenance needed. Chris is responsible for supporting a number of customers in a large metropolitan area who have a diverse range of machines (more than 50 different types/models), including some that are quite old and others that are new and state-of-the-art.

The nature of maintenance has changed considerably in the 10 years that Chris has been doing the job. Today's machines have a lot of self-diagnostic capability (via internal sensors and microchips), so they are often able to figure out what is wrong and display the problem (and remedy) to the operator. Chris is rarely called to fix the simple problems any more because they can be handled by machine operators with guidance from the machines themselves. Furthermore, many problems originate from the interaction of different machines as they pass information, so a lot of the maintenance work involves network troubleshooting from a team of different specialists.

When called to a customer site (often dispatched from a company's central monitoring site when they detect a system problem), the first thing Chris does is connect his laptop to the malfunctioning machine and run diagnostic programs that try to pinpoint the problem. Sometimes these programs can figure it out. Each machine has its own record in the customer file, including all its past maintenance history which is used by the diagnostic routines. But in many cases, the problem requires more extensive troubleshooting. Chris can then sign on to the Acme technical support network and check a database that lists all the past repair problems for that type/model of equipment (and the appropriate repairs). This database is compiled from the repair orders completed for every machine Acme services. Since specific models/types of machines often experience similar problems (due to design or manufacturing flaws), this often helps pinpoint a problem.

Finally, Chris can post a message on the technical support network about the problem that will be seen by all other technicians as well as design engineers who have expertise with a particular machine or system. As a result, Chris is likely to get many suggestions and ideas — there is a good incentive to respond because employees receive salary bonuses based on the number of successful solutions they propose. Once Chris has diagnosed the problem and ordered/replaced any faulty components, the repair report for that machine is completed which indicates all the details of the problem/repair as well as identification of sources used in the troubleshooting process. This report is automatically added to the Acme system database.

Even though Chris repairs printing machines, the job requires a high degree of computer literacy to use the various troubleshooting tools. This raises the question: Are our educational and training systems providing the kinds of computer skills that will be needed for most jobs in the 21st century?

Key Implications for Technology and Learning for the 21st Century

This chapter has presented a number of scenarios that depict how technology will impact learning and work for individuals and organizations in the 21st century. Some of the main implications include:

- Technology allows smaller groups of individuals in distributed locations to offer the same services as large organizations with centralized staffs and/or resources.
- In a technology-based environment, there is more focus on performance outcomes and less on status due to seniority, physical surroundings, or position in an organization.
- Technology will provide businesses and organizations with a significant advantage over traditional enterprises in terms of improved service and more efficient operations.
- Almost all individuals, regardless of their level or the nature of their jobs, will require high degrees of computer literacy.
- Despite the use of advanced software such as expert systems and EPSS, individuals will still need a good understanding of their business field.
- Technology improves the quality of life for most individuals, offering them more choices, faster learning, and greater control of events.

Enjoying the Future

Technology has made possible a speed and quality in the workplace that was only a dream a few years ago. In this book we have explored the tremendous capabilities and opportunities workplace technologies provide. We have seen how companies around the world have harnessed the power of technology to build a better workplace with better products and services for their customers. For those of us who can capture this power, it is truly a time to enjoy the benefits of learning and working in today's world of work.

APPENDIX A:
Glossary

Analog — transmission of information in continuous wave format (contrast with digital)

Application software — programs such as word processing, spreadsheets, databases, desktop publishing, etc.

Assistive technology — hardware/software designed to help handicapped individuals

Asynchronous — interaction or transmission of information distributed over time (contrasts with synchronous)

ATM (Asynchronous Transfer Mode) — A network transmission protocol

ATMs (Automated Transaction Machines) — One of the most successful forms of public access computing systems used for electronic banking and other retail applications

Backup — making a copy of a computer file in case the original gets damaged or lost

Browsers — programs used to access world wide web documents (e.g., Netscape Navigator, Microsoft Explorer)

Cable TV — the transmission of television signals via cable technology

CAD/CAM (Computer Aided Design/Computer Aided Manufacturing) — Software tools for the design and production of parts or structures

CBT — learning that uses computers to deliver training

CRT (Cathode Ray Tube) — The conventional display unit of televisions and desktop PCs (compare with LCD)

CD-R (Compact Disc — Recordable) — A CD-ROM drive that allows recording of CD-ROMs

CD-ROM — a format and system for recording, storing and retrieving electronic information on a compact disk that is read using an optical drive

DVD (Digital Video Disc) — A new form of Compact Disc that has enough capacity to store substantial amounts of video

Digital — transmission of information in discrete (binary) units

Electronic mail (e-mail) — the exchange of messages through computers

Electronic performance support system (EPSS) — an integrated computer application using expert systems, hypertext, embedded animation and/or hypermedia to help and guide user to perform tasks

Electronic text or publishing — the dissemination of text via electronic means

Extranet — a collaborative network that uses Internet technology to link organizations with their suppliers, customers, or other organizations that share common goals or information

FCC (Federal Communications Commission) — U.S. government agency that regulates television/telecommunication activity

FTP (File Transfer Program) — Software used to upload/download files between networks and PCs

HTML (HyperText Markup Language) — The computer language used to create documents for the world wide web

Hypertext/hypermedia — information that is connected by links

ISDN (Integrated Services Digital Network) — High capacity digital transmission service offered by telecommunication vendors

Internet — a loose confederation of computer networks around the world that are connected through several primary networks

Intranet — Internets within an organization

(LAN) Local area network — a network of computers sharing the resources of a single processor or server within a relatively small geographic area

LCD (Liquid Crystal Diode) — Flat display formats used in laptop computers and video projection devices (compare to CRT)

Multimedia — computer application that uses text, audio, animation, and/or video

RAM/ROM (Random Access Memory/Read Only Memory) — The two fundamental forms of computer memory that define the capacity of the machine

Satellite TV (also called business TV) — transmission of television signals via satellites

Scanner — device used to convert printed matter into digital form

Server — computer used to manage a network

Simulator — a device or system that replicates or imitates a real device or system

Synchronous — interaction or transmission of information in real-time (contrasts with asynchronous)

TCP/IP (Transmission Control Protocol/Internet Protocol) — protocol for information transmission over the Internet

Teleconferencing — the instantaneous exchange of audio, video or text between two or more individuals or groups at two or more locations

Television — one-way video combined with two-way audio or other electronic response systems

Virtual reality — a computer application that provides an interactive, immersive and three-dimensional learning experience through fully functional, realistic models

(WAN) wide area network — a network of computers sharing the resources of one or more processors or servers over a relatively large geographic area

Appendix B:
Internet Sites for Technology and Learning in the Workplace

The reader is encouraged to check out http://www.gwu.edu/~lto/, which will contain continuously updated web sites on the technologies and corporate exemplars contained in this book.

Distance Learning

Distance Education Clearinghouse
http://www.uwex.edu/disted/home.html

Electronic Performance Support Systems

EPSS.COM
http://www.epss.com

The EPSS Info Site Home Page
http://www.tgx.com/enhance

EPSS Resources
http://www.itech1.coe.gua.edu/EPSS

GroupWare

Collaborative Strategies
http://www.collaborate.com

Lotus
http://www.lotus.com

Intranets

The Complete Intranet Resource
http://www.intrack.com

Intranet Design Magazine
http://www.innergy.com

The Intranet Journal
http://www.intranetjournal.com

Intranet Services
http://www.iserv.co.za/infor.htm

Knowledge Engineering

Knowledge Systems Lab at Stanford University
http://www.ksl.stanford.edu

Simulation

Institute for Simulation and Training
http://www.ist.ucf.edu

Television and Video

The Business Channel
http://www.tbc.pbs.org

Federal Learning Center
http://www.fedlearn.com

Videoconferencing

World of Videoconferencing
http://www.videoconference.com

APPENDIX C:
Associations and Centers for Technology and Learning

American Center for the Study of Distance Education
http://www.cde.psu.edu/acsde

Association for Advancement of Computing in Education
http://www.curry.edschool.Virginia.EDU/aace

AECT
http://www.aect.org

American Society for Training and Development
http://www.astd.org

Center for Advanced Instructional Media
http://info.med.yale.edu/caim/C_HOME.HMTL

Center for Distance Learning Research
http://www.cdlr.tamu.edu

Federal Government Distance Learning Association
http://www.fgdla.org

Interactive Media Association
http://www.ima.org

International Centre for Distance Learning
http://www.icdl.open.ac.uk

International Multimedia Teleconferencing Consortium
http://www.imtc.org

International Society for Performance Improvement
http://www.bso.com/ispi

International Society for Technology in Education
http://www.iste.org

International Teleconferencing Association
http://www.itca.org

Office of Learning Technologies
http://www.olt-bta.hrdc-drchl.gc.ca

South East Asia Regional Computer Confederation
http://www.searcc.org

Society for Applied Learning Technology (SALT)
http://www.salt.org

U.S. Distance Learning Association
http://www.usdla.org

Index